MW00954998

THE HEART OF HOSPICE

Core Competencies for Reclaiming the Mystery

BRENDA CLARKSON

&

MYRA L. LOVVORN

TERESA! AMAZING! LOVE & LITE FOR YOUR CONTINUED JOURNEYS TO THE MYSTERY!

Myra L. Lov
June 4, 2016

ISBN: 978-1-4834-2161-2 (sc)
ISBN: 978-1-4834-2163-6 (hc)
ISBN: 978-1-4834-2162-9 (e)

Library of Congress Control Number: 2014921455

Cover Photo taken by Myra L. Lovvorn, Sunrise at OBX.

Author's Photos were taken by Gina Venn of Richmond, Virginia.

Lulu Publishing Services rev. date: 12/30/2014

Dedication

For Guy, Marc, and Lisa, whose births awakened me to the mystery and whose lives continue to teach me more than I could ever imagine. And for Clarrie Clarkson, whose death taught me much that I had yet to learn, bringing the mystery full circle. —BC

Thanks to Diane Cree Kane, LCSW, for modeling and support as I developed listening skills, mindfulness, the art of presencing, and eventually appropriate intervention. A special "etheric" thanks for the many "souls" who shared their wisdom from their deathbeds through their suffering, their stories, their tears, and their laughter. May this writing honor our journey together! Much heart-centered appreciation to the many teachers for showing me the path and encouraging me to take the first, second, and third steps so far. —MLL

Contents

Prologue: Mystery Stories

Everything is held together with stories.
That is all that is holding us together, stories and compassion.
—Barry Lopez

When I (Brenda) arrived at the nursing home for a regularly scheduled visit with Mrs. D, I was told by the facility staff that she had been unresponsive for forty-eight hours and that she was moaning constantly. As I walked down the hall, I could hear her soft moans and was struck by a sense that she was going to die during my visit. When I entered her room and let her know I was there, she made no indication that she was aware of my presence. I conducted a gentle physical assessment and found exactly what I anticipated. Mrs. D did not object to my touch nor did she respond to my words but simply continued the moaning. There was little I could do to relieve the moaning, which I took to be a sign of some sort of suffering, so I sat at her bedside and watched her breathing. I found myself singing hymns as I held her hand. I knew that church had been an important part of her life before she came to the nursing home, and she had taught me some of her favorite hymns. Gradually I realized she was gently squeezing my hand and had stopped moaning. I continued quietly singing for a while, and eventually it was time for me to pack my equipment into my nursing bag and leave to report my findings to the facility staff. Mrs. D was still quiet as I left the room. After making my report to the nursing facility staff, I returned to her room to say good-bye to Mrs. D. She was not moaning and she looked so peaceful, with a half-smile on her lips. I took her hand; she squeezed it, took two more quiet breaths, and peacefully slipped away. My premonition turned out to be accurate, my presence made a real difference to Mrs. D's dying, and I knew I had chosen this work wisely.

There are so many stories to tell. After a combined fifty-five years as hospice nurses, we could fill a book with stories!

Instead, we have decided to fill this book with things we have learned, things we have intuited, things we have observed, and opinions we have developed, which we hope will have meaning for those who choose to work with people who are dying.

Our book is written with gratitude and with the sincere hope and intention that the heart of hospice does not die in the midst of increased regulatory scrutiny, reimbursement rate cuts, accountable-care organizations, bundling of post-acute services, and any number of challenges currently facing hospice programs.

This book is our gift to those who are determined to make sure that the essence of hospice is not lost in the flurry of progress. Our book will provide the reader with the context in which our careers as hospice nurses have evolved and will describe what we have learned about the personal and professional development that we have experienced on the journey from willingness to wisdom in hospice care.

Informed by decades of administrative and clinical practice in hospice and end-of-life care and profoundly influenced by two early researchers, Bernice Harper (1977) and Joyce Zerwekh (1995), this book is grounded in a significant body of knowledge. It also includes experiential insights and tools we have developed for practical application in compassion-centric hospice programs. In returning to the fundamentals of caring for human beings at the end of their lives, we hope to validate the experiences of hospice team members who share our reverence for the mystery aspect of death while providing a developmental framework for those who are just beginning to feel the amazing grace.

Introduction

You matter because you are you. You matter until the last moment of your life, and we will do all we can, not only to help you die peacefully, but also to live until you die.
—Dame Cicely Saunders

Nowhere is the American melting pot more noticeable than in hospice. Death brings out cultural, religious, ethnic, social, spiritual, and economic variations that may be neglected or hidden in other health-care settings. Families often revert to the customs of their ancestors to provide a sense of continuity during times of loss and grief. They may call upon their personal beliefs and familiar rituals to help them through the tough times surrounding a death.

One thing we have learned in our many years as hospice nurses is that there is more to death than the simple absence of life. We have observed time and time again that regardless of the belief system of the patient, the family, or the hospice team members, something happens when a person dies that is difficult to describe, difficult to explain, and difficult to capture. As Dr. Rachel Naomi Remen points out, "care of the dying presents a bit of a quandary to those trained in the skills of physical medicine because its focus is care of the soul not just the body.... the dying are not broken and everything we cannot understand may not be awry" (2001, 338).

Death is a mystery only for the living, and it is this mystery that sets our hospice work apart from other practice settings. For us, the mystery is the ineffable transformation of spirit we have witnessed in dying patients, in grieving loved ones, and in hospice workers. The mystery shows up without invitation or expectation; it is simply a part of the human experience of death. The privilege of participating in this intimate mystery is at the heart of our practice; it is the counterbalance to the challenges that arise daily in hospice care. This mystery is at the core of everything.

Hospice nurses and doctors have developed critical skills in managing pain and other distressing symptoms. Hospice social workers counsel patients and families through the pain of grief, and hospice chaplains are skilled in identifying and managing spiritual distress. Hospice aides spend countless hours making sure hospice patients are clean and comfortable, while specially trained volunteers add to the quality of life for countless patients and caregivers. Many hospices have developed special programs using music, pets, and other complementary therapies, as well as comprehensive bereavement programs. Family members consistently evaluate their satisfaction with hospice care at rates much higher than other health-care providers. All this has led to a greater acceptance of hospice care in our communities, with the Centers for Medicare & Medicaid

Services (CMS, 2014) reporting 44 percent of the deaths of Medicare beneficiaries occurred in hospice programs in 2012.

While hospices have much to be proud of, it is essential to remember that by being invited to participate in the most important and intimate event in a dying person's life, the hospice team members become much more than "providers of health care." Working and living in a society that has lost many of the traditional comforts afforded by death rituals and practices, hospice team members have become proxies for their entire community in end-of-life care. In the presence of death, hospice team members are both witness and participant in a mystery that defines our humanness.

Our intention in writing this book is to ensure that hospice team members are well prepared for this aspect of their role. We are confident that orientation and continuing education offered by hospices prepare their team members well in the measurable skills that support safe and comfortable dying, self-determined life closure, and effective grieving (National Hospice & Palliative Care Organization, 2012). There is ample evidence that the majority of hospices prepare their staff to maintain compliance with regulations and provide safe, cost-effective care. This book is designed to overlay all these accomplishments with the core competencies needed to respect and honor the mystery surrounding death, and to reclaim the heart of hospice care.

The modern hospice movement was so called because its founder intended to fulfill the original meaning of the word by providing a way station for persons on their journey from life to death. Dame Cicely Saunders (1995) acknowledged the vital role of "meticulous nursing" and symptom management, and she also learned that there was so much more that must be done to support the personal experience of each human being journeying to a self-directed death. While we can take great comfort in the increased utilization of hospice care, hospice was never intended to be simply health care for patients with a terminal diagnosis and a certified prognosis of six months or less. Indeed, an unintended consequence of the success of integrating hospice care into the health-care system is the reemergence of the medical model at the end of life.

The degradation of the founding principles of the modern hospice movement is spearheaded by regulatory demands for evidence-based medical data as the sole criteria for hospice eligibility. At the same time, efforts for public reporting of quality measures focus on relief of pain or other physical symptoms that are amenable to quantification, while ignoring outcomes that are more difficult to measure, such as spiritual transcendence, redemption, or family unity at the deathbed. Hospice is designed to be a transformative experience for patients and families that differs significantly from the experience of care through a disease-focused model (Labyak, 2001). With its emphasis on reducing all dimensions of suffering, hospice care is so much more than a medical approach to end of life. A culture of eligibility and compliance currently permeates hospices as they face the legal, financial, and public relations challenges imposed by increased scrutiny.

However, we have written this book because we believe that now is the time for hospices to assert their unique place in our communities.

Our book reclaims the mystery of death in an attempt to counter the negative aspects of efficiency, cost-effectiveness, standardization, and regulatory compliance that dominate hospice care today.

Our book describes the core competencies necessary to bear witness to the mystery of death and to facilitate transformative experiences for hospice patients and families. At no time in our history has it been more important to reclaim the mystery that is the hallmark of the mission of hospice. So, while this book will not prepare hospice team members to survive a Joint Commission survey or help them feel more adept at responding to the demands brought on by increased regulatory scrutiny, it will bring them face-to-face with the core competencies required to reclaim the mystery surrounding death. And in so doing, it will begin to negate the current tendency toward medicalization of death that threatens the very heart of hospice care.

The following chart compares the development of hospice in the United States up to and including the model we propose, called the mystery model, characteristics of which will become clear as readers progress through this book.

Hospice Origin	Hospice Today	Mystery Model
Care model based on altruism	Business models to maximize reimbursement	Model honors mission *and* margin.
Pioneer volunteer staff	Large paid staff; volunteers as required by regulation	Development of staff and volunteers through mystery education and targeted support
Minimal regulations prior to Medicare Hospice Benefit	Excessive regulatory scrutiny for compliance with multiple sets of regulations	Regulatory compliance balanced by regard for human mystery
Patient-family needs/ strengths drive plan of care.	Financial constraints/ regulations drive plan of care.	Patient-family beliefs and values drive plan of care.
Limited resources; creative ways to meet mission through fund-raising	Diminishing resources based on concerns for margin	Model honors mission *and* margin.
Often "only game in town" to meet community needs	Competitive marketplace	Quality outcomes drive consumer choice. Mystery model hospice enhances staff satisfaction, attracting the best practitioners.

Minimal turnover; sole employer	Staff turnover rapid and frequent	Supported staff are happier; loyalty leads to reduced turnover.
Developed interdisciplinary standards for care of the dying	Little or no formal preparation related to "mystery of dying"	Core competencies for reclaiming the mystery
Dying recognized as a spiritual experience	Dying has become medicalized.	Balances disease focus with support during human mystery

The Mystery

Death is a mystery only to the living. It involves things unseen and unknown to my mortal mind that I can't even begin to comprehend.

—Ferna Lary Mills

Like many hospice workers, we came to this field in response to a heartfelt call. While the call came to us in different decades and different countries, it was the same for both of us. We were professionally embarrassed at the way people were dying in acute-care settings. Both of us silently questioned "futile" treatment and unrecognized suffering, and we instinctively knew there must be a better way. And so we came to hospice care.

Later, as we developed into competent hospice nurses, we began to realize that simply managing symptoms and attending to family dynamics was not enough. It was not enough for patients, not enough for families, and not enough to satisfy our own personal and professional needs if we were to survive in this demanding work environment. We began to ask why a patient, no longer distracted by distressing symptoms (thanks to the expertise of the hospice interdisciplinary team), may continue to suffer through an existential crisis that cannot be relieved. We asked why another refused pain medication, preferring agony to addressing his belief that death equals suffering. We asked why a mother waited to draw her final breath until all her loved ones had left the bedside. We asked why so many of our patients reported seeing deceased relatives and friends. These and other unfathomable experiences led us to a dawning recognition of the mystery surrounding death. We embarked on a never-ending journey to observe and learn to become mystery watchers in the presence of dying and death.

The mystery has been described as that which can be known only by revelation and experience and cannot be fully understood. Our own scientific background and habitual linear thinking leads us to believe that we must first be able to define the mystery that surrounds death, put words to the experience for which we do not readily have vocabulary. We should set the stage for each reader to grasp the meaning of the mystery, all the while knowing it to be unique to each person. While the Inuit peoples are reputed to have over one hundred words for snow, we struggle with only one word for the inexplicable aspects that surround a person's death.

- Our word "mystery" describes the reconciliation of body, mind, and spirit at the end of life.
- Our word "mystery" encompasses the total experience of death.

- Our word "mystery" attempts to bridge the chasm between our practice and the ineffable.
- Our word "mystery" honors the uniqueness of each person's journey toward death.
- Our word "mystery" validates the essence of each human being, for whom we grieve after each person dies.
- Our word "mystery" envelops spirit, heart, beauty, soul, and love, in an effort to be inclusive of all religions and faiths or none.
- Our word "mystery" cannot be forced or controlled; it can only flow naturally out of the experiences surrounding death.

Hospice Core Competencies

Compassionate care of the dying requires the ability to give of one's self without being destroyed in the process.
—Mary Vachon

Competence is defined as the quality of being competent; possession of the required skill, knowledge, qualification, or capacity (Dictionary.com, 2014).

Defining competencies creates an opportunity to see what it takes to be fully successful, and so the objective measurement of competencies has become the standard for performance appraisal for health-care workers. By providing a road map to superior performance, this process has merit.

Superior performance in hospice care cannot, however, be reduced to simply quantifying the measurable. The competencies that are unique to hospice, those core competencies that are the heart and soul of hospice care, are developed in the continual presence of human suffering; they are identified through witnessing the mystery of dying and death. These competencies are not easy to measure, however that must not preclude them from being valued as the very essence of compassionate care of the dying.

So much of value in hospice care is hard to capture objectively and virtually impossible to measure empirically. The value of relationships, the value of compassion, the value of open communication, and the value of human presence—none of these lend themselves to objectivity, to the checklist approach, nor even to clinical documentation.

Interdisciplinary team competencies have been described by the National Hospice and Palliative Care Organization (NHPCO), and all six competencies relate to desirable attributes and measurable qualities that are found in effective team members. NHPCO published the 2010 hospice values competencies (2012) such as teamwork, integrity, and dignity; only three of the seven competencies are patient-centered, and none reflect the human mystery of death. Nevertheless, these documents provide important structural elements that may form the basis of employee selection, orientation, professional development, and performance appraisal. Core competencies for reclaiming the mystery provide an additional dimension to such structure, rather than replacing it.

These core competencies demand much of practitioners as they accompany people on their journey at life's end. They may be taught, coached, and modeled, but to be fully understood, they must be experienced. It is only through mindful experience that the mystery at the very heart of hospice care can be incorporated into compassionate practice.

We refer to those who accept the challenge to become proficient in our identified core competencies as "mystery watchers". We chose the name "mystery watcher" to honor the profound influence the biblical phrase "Watch with me" had on the founder of St. Christopher's Hospice, Dame Cicely Saunders. She wrote that when applied to the end of a person's life, "Watch with me" demands that hospice practice must "stem from respect for the patient and very close attention to his distress" (Saunders, Baines, and Dunlop, 1995, p. 2). Implicit in Saunders's use of the verb "to watch" is taking a holistic view of the patient while continually gaining new skills in relieving distress—caring about the patient as well as caring for the patient. This means not only "listening deeply" but also providing individualized care and support to those on the journey from life to death. Mystery watchers acknowledge that this journey must ultimately be taken alone, and they recognize that "being there," being dependable, standing watch, is a gift they can give to all patients, even when they themselves may feel helpless.

In 1995, Joyce Zerwekh published research based on the stories of expert hospice nurses that confirmed our own hospice practice experiences. We began to use this study as the basis of hospice nurse orientation and training in our work in several states. Passing this research through the filter of our decades of clinical hospice practice and our experiences teaching others, we have added some competencies and modified others to define twelve core competencies that we believe are at the heart of hospice care.

We describe these competencies only as they relate to the mystery surrounding death, while respecting the skills and knowledge that all practitioners bring with them as they embark on a hospice career. Some may be highly skilled communicators, for example, and yet will need to learn additional skills as they relate to the mysterious aspects of dying and death. Becoming competent in all twelve core competencies requires continuous learning, and the rewards are often life-changing. Core competencies for reclaiming the mystery are neither for the fainthearted nor for those who are unwilling to actively participate in the personal and professional development that is intricately bound with their acquisition.

For simplicity only, we describe these core competencies as sectors in a circle spiraling toward a fully competent, wise, reflective, and profoundly compassionate practitioner. We acknowledge that our competencies often overlap, with no identified starting point and certainly no end point; the spiral of learning never ends. By superimposing this spiral onto the image of a seashell, we remind ourselves of the naturalness of death, the beauty of transcendence at the time of death, and the pearls of wisdom at the heart of the practice of hospice care.

Model of Hospice Core Competencies

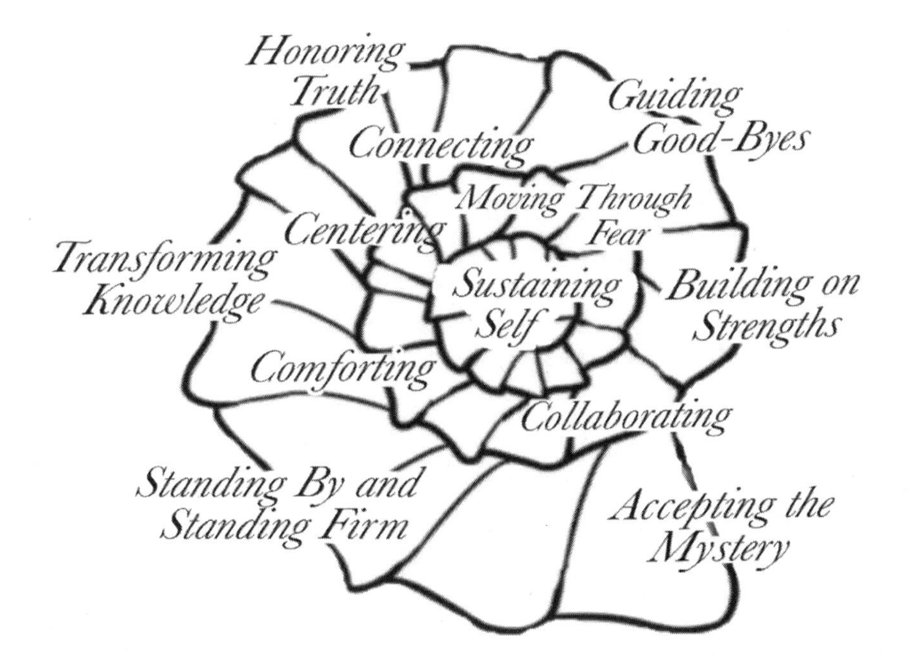

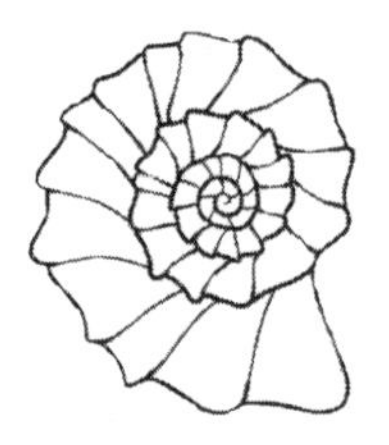

Sustaining Self

Be patient with yourself. Self-growth is tender, it is holy ground. There is no greater investment.
—Stephen Covey

To borrow conceptually from Deepak Chopra (2010), deliberate self-care is the taproot of hospice clinical practice. In the absence of strong self-sustainability, practitioners cannot fully attain the hospice core competencies no matter how technically able they may be. Compassionate care of the dying demands the capacity to give with the wholeness of being without diminishing self. With an awareness of what may trigger their own grief, hospice practitioners develop ways to care for self, maintain emotional health, and avoid being overwhelmed by both patients' and families' suffering and their own perceived impotence to always provide relief (Kearney et al., 2009).

They develop strategies to avoid any adverse effects of the ever-present grief that is woven through the hospice experience, strategies that develop equanimity and simply allow what is. They recognize that it is important to replenish self through deliberate activities designed to balance professional and personal aspects of their lives, such as exercise, music, praying, meditation, and debriefing complex cases with supportive colleagues. Staying open and healthy may be achieved by avoiding risky behaviors and by developing the ability to remain open to the suffering of others. The only true judge of self is "Self," by which we mean the authentic inner knowing of what feeds the individual's soul, the spirit, the inner-most truth of the "Self". Consequently practitioners need to develop a sense of their own personal stressors and incorporate self-care practices that are congruent with their own values and beliefs to avoid the pitfalls of compassion fatigue, burnout, or illness.

The value of healthy self-esteem has been documented for thousands of years, and its value in the development of hospice practitioners cannot be overstated. The journey to competence in hospice care is mirrored by the journey to healthy self-esteem, which is synonymous with the journey from "self" to 'authentic self" or "Self." This journey encompasses the discovery of the primacy of seeing life through one's own eyes, rather than being influenced by the perceptions and beliefs of others. Carolyn Myss, in her audio CD "Self Esteem" (2002), describes the transition from seeing the world in front of your eyes to seeing the world "behind your eyes," being in control of your life and creating experiences that challenge the spirit to grow and develop. Faced with daily exposure to loss and the suffering of others, developing a measure of self-esteem can support the journey toward understanding of the mystery of death. Without it, the practitioner may succumb to burnout, illness, and even the abandonment of hospice work before the "Self" can be realized. At first the notion of sustaining self as the cardinal competence may seem counterintuitive to the budding practitioner, who comes to hospice in order to "help other people." However, once they embrace the concept of "Self" and demonstrate the confidence to see the world differently, mystery watchers epitomize healthy self-esteem.

Hospice practitioners learn that when the heart is open, its ability to give and receive is no longer based on anything external. By staying healthily open, the capacity to give without thought of return flourishes. Receiving, and integrating the lessons learned, allows others to benefit from the hospice practitioner's experiences. Giving and receiving in complementary measure support an openheartedness that sustains the clinical practice of hospice care. Practitioners come to hospice to give of themselves, and those who succeed in this practice arena learn that mystery-based hospice care requires the reciprocal receiving of life lessons from those who are dying. Being present to multiple losses, grief, and suffering without harming self strengthens hospice practitioners, just as they are sustained and energized by rich wisdom, received freely from people grappling with the complexity and mystery of death.

Development of clear personal and professional boundaries supports the integrity of hospice practice while respectfully honoring patient and family autonomy. This leads not only to identification of safe and achievable patient-centered goals, but also to healthy professional relationships.

Additionally, the hospice practitioners' goal to remain resilient in their practice requires them to pay attention to their personal relationships, their community involvement, and their own best efforts in committing to a higher purpose; only then can the risk of compassion fatigue be truly mitigated.

Hospice practitioners are mindful of the critical importance of sustaining a healthy and competent hospice team, and so they are continually aware of stressors impacting the work life of team colleagues and accept responsibility for helping them to develop adaptive coping strategies. This practice creates and sustains a team culture in which sufficient time and resources are provided for sustainability through self-care.

Perhaps the single most effective strategy for sustaining the self in hospice is to cultivate reflective practice. Reflective practice is the systematic review of experiences from practice so that they may be described, evaluated, and subsequently used to inform and change practice (Johns & Freshwater, 2009). This is an exercise in retrospective critical thinking, colored by feelings, that helps the mystery watcher to make sense of complex experiences encountered on the journey each hospice patient makes. Reflective practice may be cathartic and insightful as patterns of experiences emerge, presenting opportunities to mold future practice based on personal learning and subsequent actions taken. Reflective practice takes time to develop and benefits from facilitation by a skilled coach, using tools such as peer feedback, reflective writing, parallel charting, and practice journaling (see chapter 5, Mystery Model Workplace, for details). When skillfully facilitated, courageous and open-minded exploration of practice by mystery watchers will:

- identify opportunities for personal and professional growth,
- celebrate strengths and accomplishments,
- develop insights into the mystery, and
- become resilient practitioners.

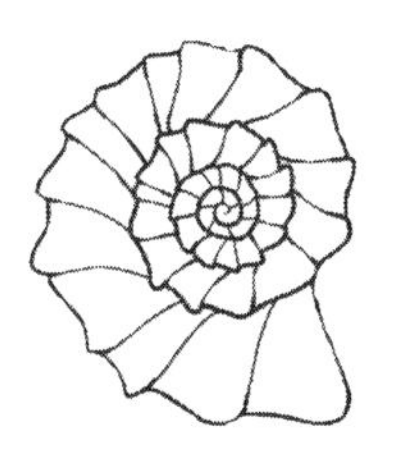

Moving Through Fear

Nothing ever goes away until it has taught us what we need to know.
—Pema Chödrön

We live in a fear-based society and receive our health care from a fear-based illness system. Fear of death is at the heart of our death-denying culture, which has resisted decades of valiant attempts to bring us to a state of remembering that throughout nature, life depends on death.

No wonder then that fears abound as patients and families face the turmoil of end of life. The courageous hospice practitioner assists apprehensive people to confront such fears, compounded now by multiple concurrent losses, about the process of dying, about death itself, and about whatever they believe may follow death.

Only practitioners who have confronted their own mortality and who acknowledge the limitations imposed by our death-denying culture can accomplish a hospice practice that excludes avoidance and denial. By learning to be fully present with each experience, hospice practitioners develop a stronger competency in facing fears related to their own mortality and the fears of patients and families also. Discerning hospice practitioners pay attention with all their senses, leading to practice patterns that transcend fear and allow compassion to emerge.

They learn that compassion comes not from cutting themselves off from the range of emotional experiences they encounter in end-of-life care, but by participating fully in these experiences (Ferrini, 1994). Through full participation, practitioners develop the courage to be fully present and vulnerable in the presence of dying and understand the wisdom in not having all the answers, along with the expediency of not attempting to "fix" all problems. Socrates reminds us that "the only true wisdom is in knowing that you know nothing," and from a stance of "knowing nothing," the enlightened practitioner is less tempted to apply a "fix."

Joan Halifax (2008) calls this "the realm of not knowing," brought about by awareness, acceptance, and presence. From this realm, the practitioner is able to address his or her own fears of perceived inadequacy, ignorance, or incompetence, as well as assisting the patient with fears related to an unexamined life, an unsatisfactory relationship with a higher power, a need for forgiveness, a fear of pain, or any of the myriad of fears that may beset a dying person.

Moving through fear demands a degree of emotional intelligence to monitor one's own feelings while recognizing the emotional states of others. Only then can hospice practitioners put aside their own fears to become fully present and accessible to people who are dying.

Moving through fear is an act of courage. Dr. Brené Brown (2010) describes "ordinary courage" as "putting our vulnerability on the line." This ordinary courage is evident in many who choose hospice at end of life, and especially in their family members and caregivers as grief takes its toll. Such ordinary courage in hospice practitioners underpins their professional development as mystery watchers. Mystery watchers are clinicians who embrace death's mystery and incorporate mystery awareness into their practice. Mystery watchers move through fear, acknowledging their own humanity and the sacred nature of their role in hospice.

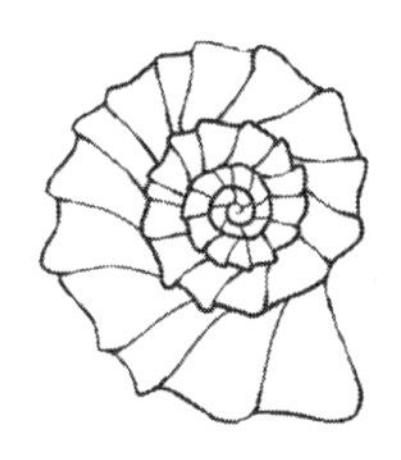

CONNECTING

When we seek for connection, we restore the world to wholeness. Our seemingly separate lives become meaningful as we discover how truly necessary we are to each other.
- Margaret Wheatley

Connecting with others is central to human well-being at any time of life, and most especially around the time of death. While it may be literally true that "we are born alone" and that "we die alone," connection helps not only humanity's evolution but also who we become as individuals (Cacioppo and Patrick, 2008). Hospice practitioners join with the patient and family to permit the caregiving process, bringing personal experiences into their practice and sometimes feeling challenged by the daily connections with dying individuals from all walks of life, religious beliefs, and cultural backgrounds. For some, spirituality is a personal connection with meaning and purpose in one's life through something greater than the self. Through religious organizations, connections can be made to share beliefs in a higher power with others in the community. At the time of death, these connections become critical for support of emotions and experiences that normalize the unseen events at death. Human beings are far more intertwined and interdependent, physiologically as well as psychologically, than our previous cultural prejudices have allowed us to acknowledge. And while social isolation can be harmful in many other settings, as people move closer to death, they may choose to limit their interactions with all but their closest connections.

The core-competent hospice practitioner recognizes the desirability of effectively connecting with

- patients and families,

- interdisciplinary team members, and
- referral sources and other community partners.

Connection has been described as the energy that exists between people when they feel heard, seen, and valued. We are hardwired for connections, according to Goleman (2006), and our connections influence our biology and our experiences. Within the context of hospice, connecting is the base from which other competencies develop. Dimensions of connecting include hearing and asking, "being there," and intentionally building trust. Affirming the patient and family as persons of value is central to developing trusting and functional relationships, connecting to enhance the well-being of both the mystery watcher and the patient/family. Some are able to do this almost without thinking, while others apply purposeful intention to establish the capacity for cooperative caring. Hospice practitioners who provide the time and space to actively listen to the patient's stories will often learn more about the family history, beliefs, and values than direct questioning could ever uncover. Stories facilitate life review and give the practitioner an opportunity to explore meaning and to be a safe sounding board for the patient when others may be unable or unwilling to participate. We are all held together by our stories, and the connections made through them endure even after objective facts are forgotten. Telling personal or family stories is not just a way to pass on information, but rather the telling creates connections of caring encompassing the patient, the family and the hospice team.

This connected community becomes a safe place to explore the mystery of dying and death, a shelter in which hospice practitioners may support the path chosen by the patient as death approaches.

Hospice practitioners intentionally build trust with patients and families by treating them with respect and dignity at all times and by committing themselves to this professional relationship. With an ongoing respect for privacy, success in effectively performing concrete tasks and responding to express needs, the relationship will prosper. Mystery watchers readily develop human connections by making sure people feel heard, seen, and valued. Other strategies for increasing trustworthiness include encouraging choices and control, consistently speaking the truth, and taking sufficient time in all interactions. Practitioners, patients, and family members alike derive sustenance and strength from such connections.

With an anticipated death as the outcome, connections with patients and families become the vehicle through which hospice care is delivered. Mystery watchers use the guidelines below to develop the skills, beliefs, and attitudes necessary to make heart-to-heart connections, spirit-to-spirit connections, and soul-to-soul connections. Such connections between the very essence of those involved become not only a crucible for the mystery, but also form the foundation of continual growth and development through the mystery watcher's own reflective practice.

Connecting: A Mystery Watcher's Guide

Intention	Strategy	Principle	Goal
Prepare self.	Establish intention.	Mystery watchers must take care of self before facilitating well-being in others.	Increase self resources through meditation, centering, breath work, and so on. Set the stage with intentionality, unconditional acceptance, and openheartedness.
Build trust.	Create a nurturing space.	Connecting is based on trusting and functional relationships.	Decrease adverse stimuli and enhance healing stimuli. Use presence, intentionality, and unconditional acceptance to maintain and nurture relationships.
Promote positive orientation.	Create a safe space.	Positive outcomes are dependent on individuals perceiving they are acceptable, respected, and worthwhile human beings.	Promote self-worth, dignity, and spiritual awareness. Maintain presence, intentionality, and unconditional acceptance to deepen the connection.
Promote perceived control.	Facilitate the story.	Perception of a degree of control is essential for ongoing personal development.	Facilitate patient's sense of intra- and interconnectedness; enhance self-awareness. Use empathy, presence, unconditional acceptance, and purposeful intentionality to create a safe environment.

Affirm and promote strengths.	Nurture growth.	The drive toward "a good death" is facilitated through respect for the individual's beliefs, values, and strengths.	Facilitate dynamic, adaptive mind-body-spirit holism. Maintain presence, intentionality, and unconditional acceptance to maintain and enhance a safe environment.
Set mutual goals directed toward transcendence.	Facilitate transformative patient-directed outcomes.	Human growth at end of life is dependent on satisfaction of lower-order needs before transformation and transcendence may be accomplished.	Maintain presence, intentionality, and unconditional acceptance to maintain and enhance a safe environment in which to facilitate life review and other communication techniques to encourage reconciliation, forgiveness, and relief of other assessed spiritual needs.

(Adapted from Erickson, 2006)

Effective connections may take time to establish, and when time is no longer available, hospice practitioners can often achieve similar results quickly just by their presence. The gift of simply "being there" is often the gift most valued by patients as death approaches; just as often, it may be the task that proves to be the most difficult to incorporate into hospice practice. The temptation to "do something" often overwhelms the most seasoned hospice practitioner. By their mindful presence, mystery watchers may bring a new dimension to the situation, a space in which solutions arise and peaceful outcomes are more likely to prevail. Mystery watchers become at one with the people they serve and with what they do, often unaware that their influence goes beyond the functions they perform. Intentionally connecting through presence may be the consummate hospice competency, turning simple work into sacred clinical practice (Tolle, 2005).

Many of the skills mastered by successfully connecting with patients and families can be applied to the vital connections hospices must make with referral sources and the communities they serve. Without such connections, access to hospice can be severely limited, depriving the dying of the many benefits of the care and services most needed.

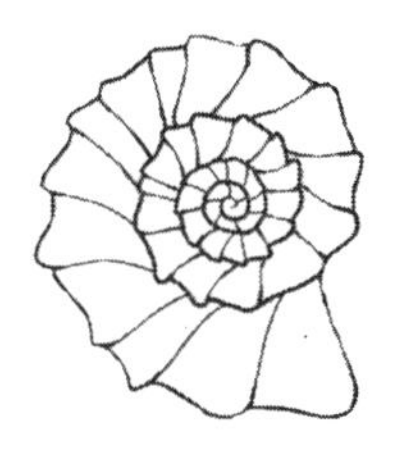

Centering

To pay attention, this is our endless and proper work.
—Mary Oliver

Centering focuses on three interwoven aspects of hospice care:

- the totality of the patient
- the unit of care
- the Self vs. self

According to the National Healthcare Disparities Report (2010), patient-centered care encompasses qualities of compassion, empathy, and responsiveness to the needs, values, and expressed wishes of the individual patient. The recent move to patient-centered care seems to be fulfilling the predication made decades ago by Dame Cicely Saunders (1995), the founder of the modern hospice movement, that mainstream health care would one day look like hospice care. Centering care on the patients' preferences, needs, and values is a way of life for hospice team members.

Advocacy for the patient's choices within the health system, the interdisciplinary hospice team, and the family system is vital for total understanding of the patient's wishes and the implications of his or her choices.

When approaching the end of a chronic life-limiting disease, many people are unprepared for the choices they may be offered by hospice and the many decisions they may be asked to make. In an attempt to discover the needs and values of each patient, well-meaning hospice team members may overwhelm both patient and family with questions and choices. And so, mystery watchers learn to mindfully balance advocating for patient autonomy with supporting a world-weary patient's decision to allow others to direct the hospice plan of care. Spiritual, legal, financial, and other personal concerns, often uncovered while simply listening to patients' stories, may be at the heart of distress for both patient and family, and as such will become an integral part of a patient-centered plan of care. Listening to stories may be a surefire way to discover authentic choices, but if not, the hospice practitioner may be called upon to facilitate difficult conversations to ensure that such choices are identified, clarified, refined, or reinforced.

Hospice has always considered the unit of care to include the patient and those who, by biology or by choice, the patient considers to be "family." Centering on the relationships involved within this unit of care enables the hospice practitioners to integrate the family's strengths and needs into a comprehensive plan of care. Often the mystery watcher is the go-between when contentious issues emerge within the family, whose dynamics may or may not change as death approaches.

Centering is a process through which hospice practitioners are able to prepare themselves for intentionally caring interventions that lead to a way of relaxing the mind at any time and especially when engaging in potentially stressful situations. Being centered not only facilitates the transformation of mere tasks into healing interactions, but it may also be instrumental in the capacity of mystery watchers to allow miracles to take place in self and others (Watson, 2008).

Mystery watchers perform best when they are able to remain relaxed and calm even in the emotional chaos that may surround a family experiencing the death of a loved one. With increasing self-awareness, the practitioner is able to mentally compact his or her energy around his or her mental or spiritual center of gravity, often referred to as "grounding" (Meditation 101: Grounding, Centering and Shielding, 2014). For many this center is physically located in the heart, for others it is in the mind, and for some it is the "seat of the soul."

Whether it is considered a spiritual practice or simply a useful technique, centering can be used quickly and unobtrusively during patient/family interactions through the practice of self-reflection, paying attention to the breath, or other meditative approaches. Mystery workers regularly practice re-centering to be in an optimal state of being at all times through longer-term approaches such as yoga, breath work, meditation, prayer, spending time in nature, and practicing anonymous service (Orloff, 2014). A longer process for re-centering involves the concept of gratitude for practitioners who are experiencing a more generalized sense of dissatisfaction. This technique endeavors to elicit a sense of gratitude independent of circumstances, making it a habit. The method includes creating personal rituals to remember gratitude and journaling about gratitude (Mead, 2014).

Centering takes practice, and when it is used well, it reminds each mystery worker that hospice gives him or her the chance to grow both personally and professionally. Centering enables hospice workers to stay healthy in body, mind, and spirit and to be fully present with patients and their families as the professional Self.

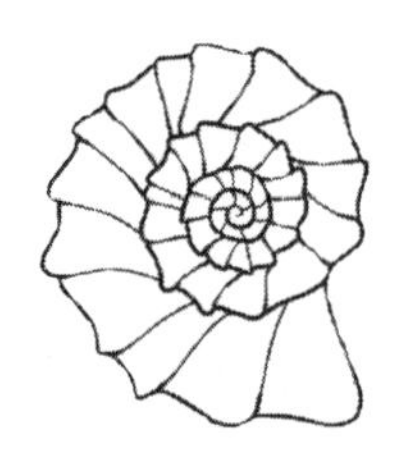

Comforting

Whether we can cure disease or relieve suffering, the most important thing we have to offer our patient is ourselves as comforter.

—William E. Cayley

The origin of "To cure sometimes, to relieve often, to comfort always" may be in question; however, the concept epitomizes the focus of hospice care. Dame Cicely Saunders recognized that discomfort is not only a physical phenomenon, but rather it affects all domains of human existence. She developed the concept of "total pain" and created a team approach to relieving all aspects of pain and discomfort when she opened the first modern-day hospice in London. Her example has been emulated all over the world as hospices strive to relieve the many discomforts related to end of life.

Relieving discomfort may involve a creative array of interventions, from sophisticated pain-management technologies to integrative therapies, to music therapy, prayer, and intentional presencing. Aggressive symptom management is the hallmark of hospice care, and practitioners are mindful of the reality that nonphysical aspects of discomfort may thwart the best efforts at pain relief. While contributing greatly to the success of hospice in enhancing the quality of life for thousands of patients, these techniques are outside the scope of our book, and we direct readers to find resources specific to their own area of interest or expertise.

In our increasingly secularized world, many live lives separated from their spirituality, and as death approaches this can be a source of deep suffering. Priorities often shift toward a need to make sense and meaning out of life as death approaches. When adapted to the hospice setting, Maslow's (1970) familiar hierarchy of needs becomes not only a theoretical model, but also the practical framework underpinning all attempts to "comfort always." Zalenski and Raspa (2006) describe their adaptations as follows:

1. pain and other distressing symptoms
2. fears for physical safety of dying or abandonment
3. affection, love, and acceptance in the face of devastating illness
4. esteem, respect, and appreciation for the person
5. self-actualization and transcendence

Relief of first-order needs is the hallmark of good hospice care and is often the first thing that comes to mind when end-of-life care choices are being made. Without attention to this level of needs, a dying person will struggle to move toward a peaceful death.

Second-order needs are for safety, both personal and societal. Hospice practitioners often hear patients say they are not afraid of death but they are afraid of dying. Mystery watchers respond with factual information to questions such as "Will I choke to death?" "What will happen to me?" "Will my family be okay without me?" Existential fears of death itself can create suffering for patients and for attentive family members.

The need to give and receive love (third-order needs) may be compromised by debility or disfigurement accompanying a terminal illness, and the desire not to die alone is often expressed as life draws to a close. With an intentional atmosphere of comforting, the mystery watcher may help to create a level of intimacy in which patient and family feel secure enough to express their innermost thoughts and feelings.

Fourth-order needs may be characterized by loss of self-esteem as patients are unable to fulfill their traditional role in the family, friendships, the workplace, and the community. Life review, acknowledging the patient's accomplishments, along with special efforts to demonstrate respect for the beliefs and values of each patient, creates comfort in the midst of sadness.

With all prior needs comforted, the dying person is more likely to achieve transcendence. It does not seem to matter whether or not the patient has had a previous experience of transcendence; whether or not he or she has pursued a spiritual practice, transcendence at end of life is a common outcome. So often the mystery transforms fear of death into clear awareness and love, and comfort transforms tragedy into grace. As with all models, people do not always follow the steps in the order outlined by the theory. Indeed, Maslow's (1970) hierarchy may become inverted as the dying person's diminishing energies are focused on existential or spiritual concerns. It is imperative, therefore, that mystery watchers be able to recognize and engage with the manifestations of spiritual distress. Often, in the safe space of an authentic and compassion-filled relationship with a hospice practitioner, patients may be encouraged to explore their unique concerns related to the universal human needs for forgiveness, reconciliation, leaving behind a legacy, and creating a sense of meaning. Mystery watchers learn that this may be the consummate way of comforting a hospice patient.

At the same time as the patient withdraws into a more reflective state, family members may be focused on the more concrete aspects of care over which they have some control, such as diet, hydration, and physical comfort. The astute practitioner forges a link between these discordant perspectives, with resulting relief of caregiver stress, anticipatory grieving, and dysfunctional family interactions. Such interventions may reduce the painful impact of dying on many aspects of family life and coping.

Spiritual matters are often avoided because providers feel unprepared to address such concerns (Balboni et al., 2012). However, by adopting a definition of spirituality such as the one offered by Dr. Brené Brown, hospice practitioners can develop a comfort level that supports their efforts to relieve spiritual distress. She says, "Spirituality is recognizing and celebrating that we are all inextricably connected to each other by a power greater than all of us, and that our connection to that power and to one another is grounded in love and compassion" (Brown, 2010, p. 64). Grounded in conscious compassion and connection, mystery watchers bring a measure of spiritual caring that transcends belief systems and religious dogma.

Joyce Zerwekh's (1995) nursing research demonstrates that relieving discomfort grows out of a strong respect for patient experience and choice. This is clearly an appropriate guideline for all hospice team members, not just nurses, in their efforts to bring comfort to patients and families. Comfort may come in many guises—a pill, a touch, a smile, a sound, a vision; the mystery watcher embraces them all.

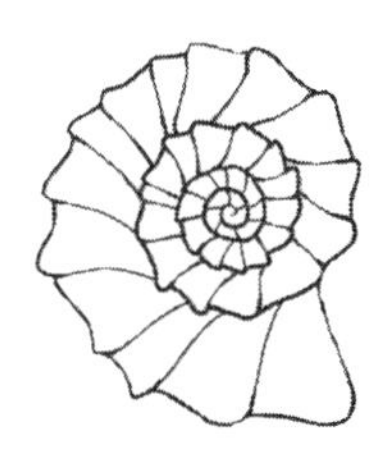

Collaborating

Alone we can do so little, together we can do so much.
—Helen Keller

Hospice practitioners collaborate in all aspects of the hospice operations; for example, by participating in the hiring process, they assess the values and attitudes of potential team members to determine how newly hired personnel will fit into the hospice team. Collaboration can be used to best effect when confronting a resistant regulatory or financial environment within the hospice program and within health care at large. Collaboration extends to other health-care providers, insurance case managers, and persons in all hospice departments to effect optimal organizational functioning.

Collaboration is also a process through which people with differing perspectives can constructively explore their differences to develop outcomes that go beyond a single limited view of possibilities. This can only happen when relationships are bound by trust and a shared value for supporting interdependencies among people. Collaboration is enhanced when the people involved are humble and trustworthy and assume good faith in their colleagues. Effective collaboration is dependent in part upon perception and insight. Deepak Chopra (2010) describes four phases of perception. To be truly insightful, looking and listening must occur on four different levels. Seeing with our eyes is only the beginning. When we look and listen fully, we involve the body, the mind, the heart, and the soul.

- body: the stage of observation and information gathering
- mind: the stage of analysis and judgment
- heart: the stage of feeling
- soul: the stage of incubation

Body, mind, heart and soul involvement only happen in a safe place, one in which relationships have been well-developed. Such a place becomes an arena for creativity and makes working together both productive and joyful. Incubation, the stage in which perception is enhanced by something greater than self, is used effectively in the collaborative practice within a hospice team. Whether it is called interdisciplinary, multidisciplinary, or transdisciplinary, mutual respect among disciplines is the cornerstone of collaborative planning for all aspects of hospice care delivery.

The hospice team collaborates to address complex patient and family issues that a single discipline could not necessarily deal with alone. Members of each discipline engage their unique knowledge base and skill set while merging and even overlapping their practice with other disciplines in order to best meet the needs of patients and families. The mystery watcher develops persuasive tactics to introduce additional team members into the care team even in the face of resistance by patients or family members. A well-crafted hospice care plan is not only a document that meets regulatory requirements, but it is also the ultimate guide to collaboration in care provision. The perceptions and brainpower of many makes light work of the plan of care! Effective collaboration is also needed with physicians and other health-care providers to ensure best practices in pain and symptom management.

Mystery watchers are keenly aware of the perceptions and insights of all who participate as patients journey toward death. By collaborating with all involved, the mystery watcher attempts to identify a mutually acceptable understanding of the mystery aspects of death. In the absence of a cohesive understanding, it is the mystery watcher who models letting go of rigid beliefs in order to honor disparate insights.

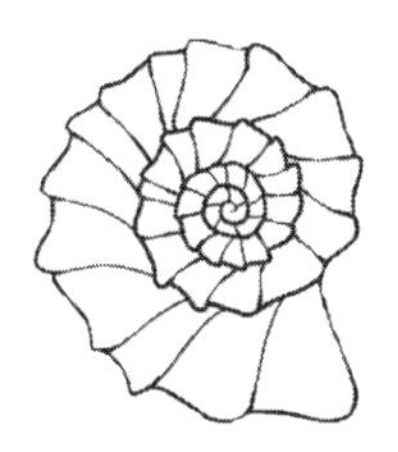

Building on Strengths

You gain strength, courage, and confidence by every experience in which you really stop to look fear in the face.
—Eleanor Roosevelt

In many health-care settings, families are considered mere appendages to patients, and often they are perceived as adding to the problems to be addressed by health-care providers rather than being seen as allies. Families have a dual role in hospice; not only do they receive care and services designed to help their healthy adaptation to an impending loss, but they also provide much of the day-to-day care for the patient. Rather than viewing family members as "problems," hospice practitioners seek out the families' strengths and build on these to improve their ability to carry out patient wishes even in the face of continuing dysfunctional family behavior. For example, Mr. B was in surgical intensive care following two months of surgical interventions with no resolution. He knew he was dying, and he wanted to go home to share his final days with his family. The staff on the unit believed his care was too complex to be delivered at home, especially as most of the care tasks would be delegated to family members. The hospice mystery watcher explored all options with Mr. B and his family, who were scared but confident that by pulling together and utilizing home hospice they could fulfill his last wish. Part of the mystery watcher's task was to collaborate with the unit staff also, to help them to see the family's strength as a valuable resource with which to meet Mr. B's needs. Once on board with the patient's desire to go home, the staff taught family members and hospice team alike all the intricacies of his care and made the necessary arrangements for a smooth transition home. His family took great pride learning all the "tricks of the trade" that they needed, and although they were individually a little apprehensive, collectively they felt confident about their ability to follow the hospice plan of care. In fact, they said they had never felt stronger than when working together to make sure Mr. B's dying wish was granted. He died comfortably two days after returning home, surrounded by his family and coordinated by hospice.

By continually assessing, validating, normalizing feelings, and encouraging adaptive behaviors, the hospice practitioner strengthens families through a time of incomparable stress. The family meeting, an often-used hospice intervention, affords the opportunity to skillfully assess family dynamics and individual and family system coping mechanisms, while also providing an opportunity for the hospice practitioner to interpret emotional and physical responses and set limits on antisocial or other negative behaviors. Death impacts all family members and all aspects of family life to a greater or lesser degree, but when families are supported and strengthened, they often go on to exceed their own expectations in delivering care to a family member. This success becomes a vital component of their subsequent ability to cope with the inevitable loss.

When facing death, many people are surprised to discover strengths that have not been apparent in earlier life. Hospice practitioners are ideally placed to skillfully uncover such strengths and to help the person build on these as death approaches. Out of the fullness of their own being, mystery watchers can connect with patients to discover strengths in the totality of the patients' lives, rather than simply from the limitations inherent in their current status as persons with a terminal diagnosis. For example, mystery watchers may be asked to facilitate reconciliation with family members that the hospice patient would not have been able to contemplate if such strength had not been revealed.

Building on the strengths of the team is an important contribution that the hospice practitioner can make to ensure that the most appropriate team member responds to the needs of patients and families. They invest in the success of new team members, nurturing confidence and competence as clinical judgment develops. Hospice practitioners validate the growing skills of colleagues, teach and model desired competencies, and provide emotional support during stressful times. Mystery watchers provide clinical leadership by acting as coaches or mentors and are active participants in staff support activities. By modeling respectful collegial relationships and demonstrating a positive attitude to the hospice program, hospice practitioners build the strengths of the clinical team and build the capacity of the total hospice organization.

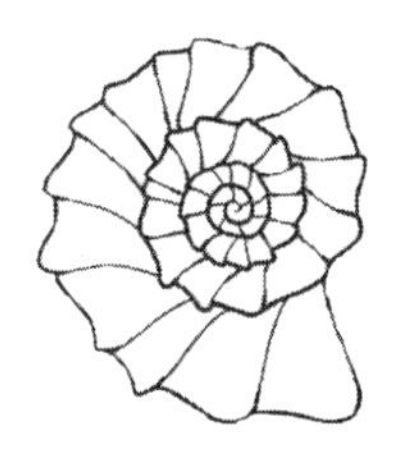

Guiding Good-byes

Only in the agony of parting do we look into the depth of love.
—George Eliot

Hospice patients have typically experienced many losses because of their terminal disease and are anticipating additional multiple losses as they leave behind family and friends. Hospice practitioners sensitively explore such losses while transforming patients' hopes and goals into more achievable outcomes. Reframing, validating, and actively attempting to resolve any aspects of "unfinished business" are interventions that the hospice practitioner may find invaluable as a patient's death approaches. Hospice practitioners often witness regrets or the need for forgiveness, and are called upon to facilitate the patient's process of forgiveness of self and others. If the process is to have real meaning for all concerned, it needs to go much deeper than offering resolution through a quick-fix option of encouraging apologies or making contact with long-lost relatives. While this may allow people to feel better about a specific concern, it is usually short-lived. For mystery watchers, listening to the story is the first step, and assisting the person to acknowledge the harm that was done, to feel worthy of being forgiven, and to fully embrace the experience of the other who was involved demonstrate that forgiveness goes far beyond a simple pardon (Ferrell, 2012). This level of forgiveness can be transcendent for the patient and life changing for survivors.

Teaching the signs and symptoms of impending death and normalizing experiences of death are ways in which the hospice practitioner can reduce the fears of both patient and family. With fears minimized, families develop confidence in their abilities to meet the patient's needs while embracing the mystery surrounding the death. Families are often amazed at their own success as caregivers, taking great comfort in knowing they have "done everything we could." This is the basis of safe and effective grieving, which follows the loss. They often find strength in retelling family stories, recalling happier times, and celebrating the accomplishments of the person who is dying. Mystery watchers often develop an intuitive sense for predicting or interpreting events surrounding death, which provides a comfort to family members.

Dispelling any fear the patient may have of abandonment, modeling "leave taking," and appropriate language can be one of the most precious gifts a hospice practitioner gives during the lifetime of the patient. The final good-bye for family members may be made easier if the practitioner guides them in appropriate postmortem care or assists them with practical issues such as funeral planning.

As practitioners say good-bye to many patienrs and families, it is important for them to appropriately reflect on the care and services provided by hospice, acting on opportunities for improvement when necessary. And acting on opportunities for personal growth through lessons received from the human mystery of dying and death. Mystery watchers learn with every good-bye.

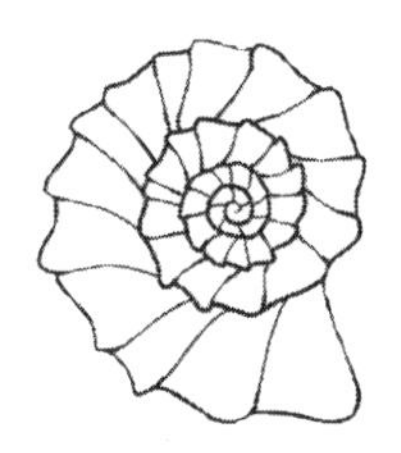

Honoring Truth

It takes two to speak the truth;
one to speak, and another to hear.
—David Henry Thoreau

Truth in the complex and multidimensional context of hospice care is often difficult to discern. Hospice practitioners recognize that statements thought to be truths correspond to the reality focus of the speaker and that such perceptions may transition over time.

They also learn that seeking the truth requires looking within at their own beliefs and intentions before embarking on exploring external truths. In order to see things honestly, hospice practitioners must understand their own emotional investments in a given situation, as well as the situation's external appearance. Both inner and outer realities must be acknowledged. Speaking the truth about real issues in real peoples' lives requires a considerable knowledge base and a confidence in the ability to recognize truth. With their genuine concern for others, mystery workers promote healthy interactions among all those concerned with aspects of the truth. Mystery workers learn that painful truths only emerge in the presence of compassion.

Telling the truth is considered by many to be a moral obligation, and according to Zerwehk, it includes asking difficult questions, saying difficult words, and facing the normal human tendency to avoid and deny (1995). Hospice practitioners become adept at knowing how far to push against avoidance and when to back off, respecting the legitimate choice of denial as a coping mechanism. The mystery worker may therefore temper the obligation to tell the truth by engaging in active listening, observing behaviors and nonverbal cues, and avoiding assumptions and defensiveness before attempting any caring disclosures.

Timing is a vitally important consideration when planning interventions related to truth telling, as is inclusivity, leading the hospice practitioner to ensure that all appropriate caregivers and family members hear "the same thing at the same time." Joint visits (with other hospice team members) and family conferences are invaluable strategies.

Many families have long-held secrets, and at the time of death, there is often the impulse to bring things out into the open and tell the truth. Honoring such a truth can only be accomplished as part of a trusting relationship and only when all those involved feel safe and secure in expressing their innermost thoughts. Mystery watchers make fine catalysts for truth telling.

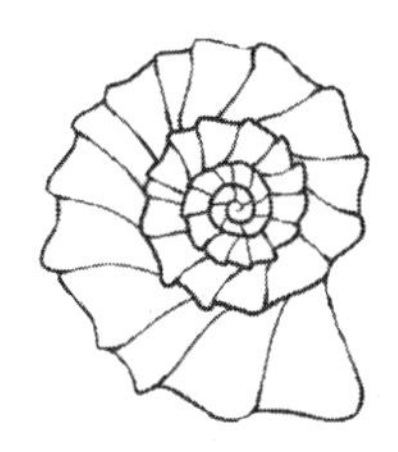

Transforming Knowledge

Knowledge alone does not produce wisdom. Transforming knowledge into wisdom requires input from the heart.
—Daisaku Ikeda

When embarking on a hospice career, practitioners bring with them knowledge acquired through formal and informal learning filtered through their beliefs and values and molded by their own aspirations. The new hospice practitioner comes to understand that knowledge alone will not assure him or her of success in end-of-life care. Success takes many forms, and a "good death" may look nothing like the "perfect death" practitioners had hoped to participate in when they first took up this work. It is through successes and challenges and their chronic exposure to human suffering that this knowledge transforms to knowing.

Knowledge moves from simply being a function of the mind to being a function of the heart and soul, a knowing that integrates the entire being. This knowing allows practitioners to no longer engage simply from a perspective of observation and analysis, but from a feeling place where profound insights are possible.

As knowledge is transformed, they experience the two basic components of self-esteem:

1. They trust their mind to respond effectively to challenges, and
2. They are confident that success and achievement are appropriate for themselves.

As healthy self-esteem grows, hospice practitioners are able to develop the skill of intuition (Myss, 2002). This kind of knowledge, the intuitive kind, can help mystery watchers to interpret the extraordinary and be comfortable in the sacred space that often surrounds death. Hospice practitioners respect and honor the opportunity to learn from other human beings who are approaching the transition from life to death. Knowledge implies having answers; knowing understands the wisdom to be found in uncertainty and of being open to the notion that this hospice practice may bring them far more than anyone can understand. The paradox of knowing is that it is unteachable.

Mystery watchers make every effort to ensure that knowledge transformation becomes the goal of all team members through strategic storytelling and modeling presence and by intuitively finding the right way of explaining the unteachable, thus making the ineffable mystery more accessible to the learner.

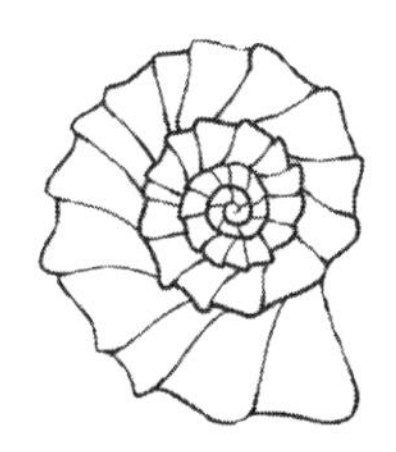

Standing By and Standing Firm

Be sure to put your feet in the right place, and then stand firm.
—Abraham Lincoln

The hospice practitioner is often called upon to stand by when choices are being made, when pains are being endured, when hopes and fears appear to be lost. By letting go of their own agenda and bypassing their own ego, hospice practitioners can provide a safe and secure environment in which dilemmas and discomforts can be explored and ultimately resolved. Hospice practitioners understand that every patient encounter is a unique event in the experience of the dying patient and that by using their whole being to influence their interventions and their presence, the best outcome is possible. Being a witness to the death of another strengthens the human capacity of each hospice practitioner and provides clarity of purpose for subsequent death vigils.

Standing by means making the most of opportunities to give and receive life lessons in the intimacy that surrounds dying. "I see you; I hear you; and what you say matters to me" (Winfrey, 2011) becomes the mantra of the mystery watcher as a stalwart and effective advocate.

Hospice practitioners are often called upon to stand firm and be "the calm in a storm." The storm may be brought about by fears and confusion as a person approaches death, or by the maladaptive coping of a family at the bedside. The storm may arise at an interdisciplinary team meeting when team members are at cross-purposes in developing a hospice plan of care. The storm may be caused by a breach in customer service, with the customer being an attending physician, a hospital discharge planner, a grieving survivor, an insurance case manager, or anyone with a connection to the hospice program.

I (Myra) want to share a story with you. Toward the end of our interdisciplinary team meeting, a hospice inpatient unit nurse called me out of the meeting to let me know that our patient, who had recently transferred from ICU, had died and his son had been the one to find him. I was expecting Mr. T; however, I was unaware that he had actually been transferred. The hospice and palliative care fellow accompanied me to the room to meet with Mr. T, Jr. As we entered the room, I walked to the bed, glancing at the patient, and then I noticed a very husky, well-dressed gentleman who turned from the open window to face me. Before I could formulate what I might say, he began to express his anger as he moved forcefully toward me with the bed between us. He shared that he was a minister; he knew a lot about death and he knew about suffering. "I promised my father he would not suffer, and now he has. He died alone! I did not keep my promises to him," he shouted. Time stands still in many different experiences in one's life, and this was one such moment. It seemed like he would never stop shouting his anger in my direction, though it was only for a few minutes in reality.

When he finally stopped, I calmly said, "I will leave you with your father; let us know when you are ready for us to come back in." As I turned to walk out, the fellow was standing with her back against the door, hand on the knob, and tears were flowing down her face. This inexperienced doctor was overwhelmed by the energy of the anger expressed. She would have been unlikely to formulate a response in the absence of my mystery-watcher expertise. As we talked outside Mr. T's room, she reflected on this emotionally draining situation and began to understand that the anger could only create a problem for her if she "took it on." She was still nervous that the son would blame hospice for his father's "suffering."

After debriefing with the team, I left the nurse's station to begin the process of postmortem care when Mr. T, Jr. came out of the room. His face was soft, his eyes clear, and he walked directly toward me. We stopped and he looked deeply into my eyes, connecting at a spirit level, and I "knew" that he was now at peace with the way his father had died. He apologized and thanked me for everything that was done for his father. I was surprised by the apology for his behavior and even more surprised when he asked for a hug. In this moment, the power of connecting normalized his raw grief, and my ability to stand firm in the anger storm was the vehicle through which this healing came about. He would still need to reflect much more on the circumstances of his father's death, his own feeling of guilt, and ways he could help his family cope with this profound loss. But he had made a great start toward understanding the power of the mystery.

In standing firm, hospice practitioners are able to intuitively conduct a root-cause analysis to discover the origins of the storm. They transform their knowledge about the storm into understanding the dynamics behind the human behaviors that are causing the storm. The mystery watcher intentionally diffuses the energy of the storm by standing firm and allowing the force of that energy to dissipate. For example, in the face of a patient's anger, hospice practitioners acknowledge the deeply humanizing effect of grief and its expression, often in the form of anger. By valuing grief and supporting people through the experience of grief, hospice practitioners stand firm inside the human drama. Whatever the underlying dynamic behind a display of anger, mystery workers purposefully allow themselves to be a container for the anger to avoid it spilling over into harmful behaviors. Hospice practitioners learn how to hold the anger without absorbing it by taking it personally. By standing firm, mystery watchers may provide language or demonstrate behaviors that patients and families can adopt as a bridge between concrete here and now and the developing mystery as death approaches.

When standing by, the mystery watcher is honored to witness, without judgment, the unique process of each death, while learning from each unfolding mystery.

In standing firm, mystery watchers remain emotionally intact, self-sustained, and strong for the next challenge.

Standing by and standing firm cultivates a resilient spirit, an ability to overcome adversity, and a desire to continue the journey as mystery watcher.

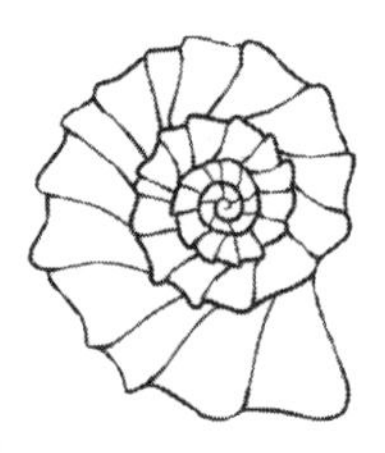

Accepting the Mystery

Witnessing the transformation of the dying process with any degree of receptivity, participating as a fellow human being enters deeply into grace, is unforgettable and inexpressible.
—Kathleen Dowling Singh

Hospice practitioners develop a new kind of knowing that embraces the human mystery of death. Mystery watchers confirm the centuries-old wisdom of dying as a deeply spiritual process, not simply a medical event. They are at the forefront of an emerging willingness on the part of the public to explore death, to face mortality, and to "allow the mystery to enter our being" (Dowling Singh, 1998). They begin to understand that this mystery cannot be forced or controlled; it can only flow naturally out of the experience of death, which is unique to each human being. The knowledge that death is defined by an absence of cardiac and respiratory functions is transformed by observing, over time, phenomena that often accompany the actively dying stage, such as

- the control the dying often exert over the exact time and circumstances of their death,
- the compulsion loved ones may feel to be present at the death,
- nearing death awareness, and
- the grace and tranquility that often accompany death even in the most chaotic environment.

Hospice practitioners understand that patients' needs are now totally focused on the huge life transition that faces them. Efforts to provide physical and emotional comfort are now the patients' secondary considerations.

Mystery watchers are often the first to recognize the language of metaphors that often accompany the final hours or days of a person's life. They begin to affirm, for example, a patient's expressed desire to "go home" or to prepare for a journey as indicators of an imminent death. Nearing death awareness takes many forms, and yet there are commonalities that the mystery watcher comes to understand, such as deathbed visions of deceased family members or religious figures (Callanan and Kelley, 1992). Teaching family members about the spiritual transformations that may be happening when patients exhibit terminal restlessness, agitation, or lucidity is a compassionate intervention aimed at enabling them to acknowledge the mystery unfolding as death approaches (Pearson, 2004).

By acknowledging the grace of death, hospice practitioners can positively influence the dying experience for both patients and families. Regardless of their own discipline, they feel empowered to provide whatever the patient and family need. If they can explain the unexplainable, normalize feelings, and teach at least one person present to hold the sacred space of dying, a mystery watcher may choose to leave the patient's home in order to allow the dying person and the family to experience this tender, private, and meaningful event together, without the intrusion of hospice.

Mystery watchers also know when it is expedient to intervene by remaining present as a participant-observer and coach in the transitional process of dying. Being part of this mystery is personally and professionally rewarding; hospice practitioners are self-affirmed without taking credit for outcomes. Being part of the mystery feeds the soul of hospice practitioners, sustaining not only their practice but also their whole being, which enables them to continue in the sacred work of patient and family care and the rewarding work of mentoring newer hospice workers.

Nurse theorist Jean Watson recalls a quotation hanging on a wall in India: "Life is not a problem to be solved, but a mystery to be lived" by M. Scott Peck (1978), which expressed some of Watson's own conceptual concerns about caring. And so it is with death. Our efforts to improve end-of-life care will falter and our humanity will be diminished as long as we approach death as a problem to be solved rather than a mystery to be experienced.

The mystery watcher understands the folly in defining hospice quality by simply valuing empirical evidence, while at the same time negating the human need for wholeness, mystery, discovery, and transcendence.

The Mystery Path

Wherever you go, go with all your heart.
—Confucius

We have both spent many hours attempting to foster a continuous process of knowledge discovery, integration, application, and reflection in hospice practitioners. Rather than simply encouraging hospice team members (and ourselves) to accumulate more and more knowledge, we have tried to expand in them (and ourselves) the capacity to provide superior hospice care through transformational education and through its application in real-life situations.

Before the Medicare hospice benefit was enacted, before documentation became hospices' holy grail, and before the medicalization of hospice care, many of us learned how to do the work by simply doing the work. There were no experts; we were developing the body of knowledge as we developed our hospice programs. There were no orientation manuals or guidelines, so many of us learned what to expect as we developed as hospice workers from research conducted by Bernice Harper, PhD, MSW (1977). Now we find that few hospice clinicians have heard of this work, which led us to develop and teach our own adaptation of this important study based on our collective insights and experiences.

We call it the "mystery path" in recognition of the multiple mysteries that surround death *and* hospice practitioners' personal and professional development as they learn how to flourish in their chosen field. We call those who journey toward this wisdom, mystery watchers.

The mystery-path model describes systematic phases that influence the development of hospice practitioners, and its description on paper makes the path appear much more precise than is actually possible in reality. Practitioners do not necessarily pass from phase to phase sequentially, and what looks like a linear process may in actuality be a much more circuitous journey. Experience is a prerequisite to competence, and so situations arising along the way may accelerate or slow down a person's development.

The skills-acquisition model developed by Stuart E. Dreyfus (1981) and adapted to nursing by Benner (1984) describes three aspects of skilled performance plus five levels through which the learner moves in acquiring such skills. Respectful of their model, which describes the movement from reliance on abstract principles to the use of concrete experiences and finally developing from detached observer to involved performer, our hospice experience informs the design of our mystery path. Very few people embark on a career in hospice practice without already demonstrating considerable clinical skills or without some ability to apply contextual meaning to aspects of the dying process, and so we describe only four phases through which all mystery watchers pass. We have also intentionally chosen titles for each phase that reflect our understanding that the art of mystery watching is as important as the science behind good clinical hospice care.

The mystery-path model depends on peers, mentors, coaches, and supervisors to guide the new employee into professional maturity. Too often, however, those tasked to provide this guidance have themselves been poorly prepared in core competencies for reclaiming the mystery. This leads to a distortion in the time it takes to move through the model and may lead to systemic frustration within the hospice organization. It is possible for mystery watchers to see themselves in several phases at once, and we have found this to be quite normal. It is also common for hospice practitioners to get "stuck," taking years to pass through a particular developmental phase without the structure of the mystery-path model.

The mystery-path model has reduced a complex process into a manageable system that can also be used to inform management practices, performance appraisal, rewards and compensation, workplace design, and staff support.

We hope that mystery watchers, by understanding the mystery-path model, will become their own coaches, and by recognizing developmental characteristics in colleagues, they will be able to help their teammates.

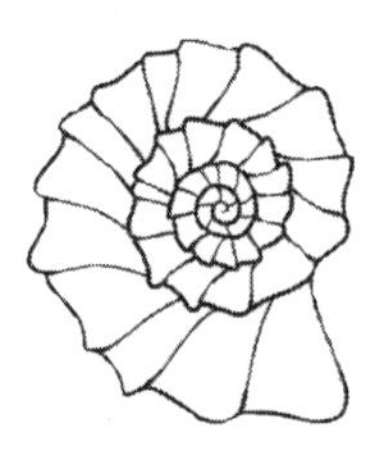

Phase 1: The Novice

A novice is defined as a beginner; one who is not very familiar with a particular subject, one who has no experience of the situation in which he or she is expected to perform. The performance of the hospice novice is characterized by intellectualization, independent practice, concern about unfamiliar work processes, and emerging awareness of the true nature of hospice, often leading to death-related anxiety and discomfort.

General Features

Novices may demonstrate acceptable performance when judgments and actions are based on previous experience and prefer to practice by adhering to the new rules and standards established by the new practice setting. As they acquire and develop new skills, they begin to appreciate complex situations; however, they are only able to achieve partial resolution without guidance; novices will still need supervision for the overall task. When entering a new practice setting or if performance goals and tools of patient care are unfamiliar, a practitioner may revert to the novice level as context-dependent skills can only be acquired in real situations (Benner, 1984).

Hospice-Specific Features

New hospice professionals have much to learn about the structure and processes of end-of-life care. Hospice novices, although they may be competent beginners (based on previous practice settings), exhibit some degree of anxiety about new policies and procedures, fundamentals of pain and symptom management, and other concrete issues. Novices take comfort in the safety of dealing with problems through the implementation of standards of practice they bring with them from work in previous settings. Their rules-based practice patterns lead to rigid thinking and limited creativity.

Novices are usually agreeable employees, who superficially accept the demands of their new role although they may exhibit a degree of inflexibility in the workplace. At this beginning phase, novices prefer to obey the rules and rely on structure and standards to guide them. They often feel "called" to do this work and are very eager to please, and might even accept a challenge for which they are not yet prepared, leading to disappointing outcomes.

Novices, focusing on learning and doing, may adopt a servant attitude and feel responsible for all aspects of hospice patients' outcomes, even knowing that the interdisciplinary team is there too for support. While eager to earn their respect, novices have not yet become comfortable accepting help from others.

By concentrating on tangible services, novices avoid emotional involvement with the grief and pain that surrounds this new practice setting. While attempting to stay emotionally safe inside their professional demeanor, they gradually begin to feel the sadness inherent in continued exposure to dying and death. The honeymoon phase of working in hospice begins to feel uncomfortable, and they may feel guilty for questioning their decision to work in hospice. It is often less painful to focus on families, rather than risk exposure to the suffering of the patient who is dying; therefore, novices often spend less time with patients and more with family members. They may initiate conversations that are factual and knowledgeable, even philosophical, but not personal. Such conversations protect novices from their own feelings. As they accumulate new and useful knowledge, they also demonstrate a level of discomfort based on an emerging awareness of the emotional toll on patients/clients, families, and themselves. And as this awareness unfolds, hospice novices may avoid confronting families, may be unable to engage with patients who want to talk about their experiences of dying, and may withdraw from other team members in an attempt to hide their real feelings (Harper, 1977).

Novices spend a great deal of time "in their heads," acquiring knowledge and learning the intricacies of the new practice setting. This tends to insulate them from feeling overwhelmed by dying and death. Anxiety may build as they experience more emotional trauma, and this may play out in over identification with the patient, with illnesses, and with self-doubt.

Truly successful students of the mystery surrounding death have a "beginner's mind," which allows them to suspend what they already "know" in favor of the learning that comes from continual exposure to dying and death. Navigating the novice phase requires conviction and commitment on the part of the novice *and* the hospice organization. Developing skills to sustain self, systematic coaching, consistent feedback, and a supportive work environment conducive to open communication lead to success.

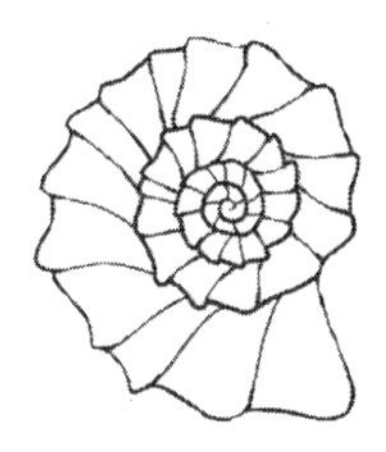

Phase 2: The Apprentice

An apprentice is defined as participating in a system of training to develop a new generation of practitioners of a structured competency-based set of skills. The performance of an apprentice may vacillate between that of an advanced beginner, with intermittent practice at competent or proficient levels. The apprentice may not yet have consistency in performance and may need continual guidance from performance coaches, mentors, and peers. Characteristics of the hospice apprentice phase include unanticipated emotional self-awareness, increased discomfort, codependence, and awe as the wisdom of this developmental phase emerges.

General Features

Apprentices develop a good working knowledge within their area of practice, and they are able to demonstrate efficiency and are often coordinated and confident in their actions. They begin to rely on their own judgment and cope with complex situations through deliberate analysis and planning. With sufficient real, meaningful, situational exposure, apprentices begin to recognize clinical patterns and overall care characteristics, to which they appropriately apply previously gained skills. Apprentices begin to see their actions as part of a longer-term approach, and goal-focused care is usually completed within acceptable timeframes (Benner, 1984). Apprentices progress along the path toward autonomous functioning and only intermittently require close supervision and guidance.

Hospice-Specific Features

Hospice apprentices develop an increased knowledge base; however, this does not necessarily lead to an increased comfort level within the newly chosen field of practice. Hospice apprentices become increasingly frustrated and even guilty about their own robust health in the face of a dying patient's weakened state. The reality of the sheer number of patient deaths and the associated suffering generate a need to confront their own mortality, leading to a traumatized state of awareness (Harper, 1977).

If the hospice program encourages safe and supportive exploration of such discomfort, apprentices will feel validated and affirmed, gradually transforming the discomfort to an openheartedness in their practice. Openheartedness is not a technique, but an emotional willingness that moves the apprentice beyond mere intellectual definitions of death and dying to deepening emotional and spiritual awareness. Parker J. Palmer (2011) tells us that the heart is a word used to describe a larger way of knowing, of receiving, and of reflecting on our own experience that takes us deeper than the mind alone can take us. It is through this larger way of knowing that each death opens the hospice apprentice's heart with less and less fear of the changes being observed in self and others, leading to a heightened understanding of the magnitude of hospice practice. By moving from the mind center of practice to the heart center, the apprentice begins to simply "know" how to respond to difficult situations and begins to trust this other way of knowing.

Hospice apprentices start to value intuition, the direct way of knowing, the effortless perception that connects them via the body to everything on the physical plane and through the heart to their own soul and the souls of others. By actively cultivating their intuition, hospice apprentices expand their skills and develop their emotional resilience.

Hospice apprentices begin to feel the satisfaction of knowing so much more about their role than they did in earlier days as a hospice team member. They relish the ability to grow, analyze, plan, and intervene to resolve ever more demanding clinical situations "single-handedly." They may not yet fully embrace the team concept of care even when challenged by the complexity of care demands and the emotional toll of daily exposure to human suffering. Rather than ask for help from team members, hospice apprentices tend to rely on their own ability to cope with all challenges.

As newfound openheartedness enhances their interactions with patients and families, it also allows the pain of mourning and grief to deeply affect mystery watchers. As the capacity to "feel" grows, so too does the emotional pain. Obliged now, by these painful emotions, to confront their own mortality, often for the first time, mystery watchers are faced with spiritual and existential distress. Coping with these work-related stressors at a time when, paradoxically, the practitioner may be experiencing successful outcomes in other aspects of his or her work, leads the mystery watcher into the "grow or go" phase that seems to be a universal phenomenon in the growth and development of hospice professionals. The discomfort of this dilemma demands a decision. "Should I go or should I stay?" "This is just so hard." "Do I want to continue this painful work?"

Success in this phase requires much courage and effort in dealing with soul-searching questions that deserve reflective answers. If hospice apprentices are assured of the hospice program's commitment to supporting them through this difficult phase, they will be motivated to face their feelings and anxieties, move fully into a heart-centered practice, and continue into the next phase of the mystery path. Through this synergy and a decision "to grow," apprentices find their true place on the team. If the decision is made "to go," the hospice program must honor the decision and assist apprentices to move on to a more suitable practice setting. With appropriate coaching, support, and guidance, apprentices can move forward in their careers elsewhere with personal and professional integrity intact (Harper, 1977).

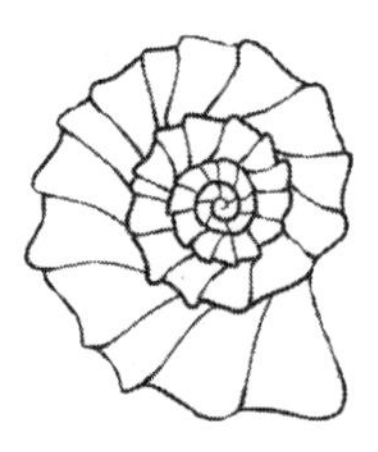

Phase 3: The Artisan

An artisan is defined as a person engaged in or occupied by the practice of a craft, who may through experience and talent reach the expressive levels of an art. Building emotional fortitude, embracing interdisciplinary team practice, and developing deep compassion characterize the performance of the hospice artisan.

General Features

Artisans have developed an understanding of their discipline and area of practice. They routinely achieve acceptable standards, taking full responsibility for their work and the work of others when applicable. Artisans perceive each situation as a whole, rather than in terms of parts or aspects, understanding overarching implications and longer-term goals (Benner, 1984). Having truly learned from experience, the artisan uses a holistic understanding of situations and problems. There is a willingness to modify plans when faced with the unexpected; decision making soon becomes less labored as analytical thinking becomes more confident. The artisan begins to acknowledge the other ways of knowing, such as an inner voice or intuition.

Hospice-Specific Features

Hospice artisans resolve many of the heartfelt questions that belabored the "grow or go" phase and make a commitment to develop both personally and professionally in the interest of providing optimal end-of-life care. They enjoy a sense of freedom from their previous discomforts and preoccupation with their own mortality. This freedom is expressed in more creative, soulful, and less rule-bound approaches to problem solving.

Artisans are increasingly aware of circumstances and emotions impacting their daily work life, feeling the same pains as before, but now they have developed the resilience to recover and remain effective in spite of any residual sadness. They become adept at intentional responses, providing emotional support to patients and families while carrying out their clinical and administrative tasks. Hospice artisans exhibit the emotional intelligence needed to monitor their own feelings, while recognizing the emotional states of others. They are able to be fully present with each patient and family, even when grieving for their own most recent losses. It is this very presence through which they develop the sensitivity to grieve and the resilience to recover.

Hospice artisans are beginning to recognize that intellectualizing in the face of death may simply be a way of avoiding pain. Increasing self-awareness leads to the development of a stronger competency in facing fear, of looking within for answers, and for actively involving other disciplines in problem solving.

Artisans are cognizant of their limitations and begin to develop effective strategies to meet the paradoxical challenges inherent in compassionate patient care coupled with masterful self-care. They are beginning to control their practice, to support and educate their colleagues, and to truly involve all team members appropriately in the provision of holistic, humane, and culturally competent care. This phase of hospice practice is demonstrated by honesty, humility, and rededication to personal and professional self-mastery.

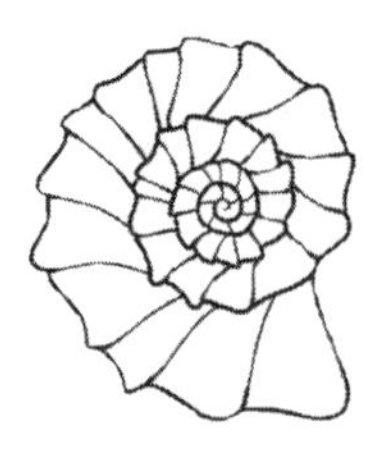

PHASE 4: THE SAGE/MASTER/ELDER

A sage/master/elder is defined as a person with extensive knowledge and ability based on research and experience of occupation. The hospice sage/master/elder is characterized by the emergence of profound wisdom, transformation, interdependency, and courageous compassion.

General Features

Sages/Masters/Elders demonstrate authoritative knowledge of their chosen discipline and deep understanding of their area of practice, and they achieve excellent standards of work with relative ease. With a holistic grasp of complex situations, sages/masters/elders move seamlessly between analytical and intuitive approaches, going beyond existing standards and creating their own interpretations (Dreyfus, 1981). With an ever-expanding background of experience and an intuitive grasp on situations, the sage/master/elder is able to act swiftly and accurately without wasteful considerations of alternatives (Benner, 1984).

Hospice-Specific Features

There is no end point on the mystery path. For the purposes of our model, however, we are calling this the era of the master, the sage, and the elder. This is the phase during which mystery watchers continue to hone their craft, accumulating valuable expertise and exquisite empathy.

- As sage, the mystery watcher's practice is characterized by wisdom, prudence, and good judgment.
- As master, the mystery watcher becomes not only a meticulous clinician but also a transmitter of knowledge, showing thoughtful

leadership and resilience in the face of clinical and workplace challenges.

- As elder, the mystery watcher develops authority by virtue of experience, becoming a repository of cultural and philosophical knowledge related to end-of-life care.

Though the journey may have been traumatic and anxiety-provoking and even though it may have challenged long-held beliefs and values, the hospice sage/master/elder practitioners have the feeling of "coming home" in their chosen practice arena.

Sages/Masters/Elders have mastered the art of fully giving self to other, respecting the dignity of patients and families while retaining personal and professional integrity. These mystery watchers transform knowledge into wisdom. Their deep compassion transforms interactions with patients and families, with colleagues, and with the Self. Sage/Master/Elder practitioners find the journey itself to be rewarding and are at peace with the unanswerable questions that emerge during the mystery of dying and death.

Sage/Master/Elder mystery watchers have acquired a great deal of knowledge, organized in ways that reflect deep understanding and that support mature situational discrimination. And yet far from feeling as though they have all the answers, sage/master/elder mystery watchers are comfortable with uncertainty in the face of dying and death. They understand the gift of presence, and they understand the gift of mystery.

Sage/Master/Elder mystery watchers willingly use their whole being in service to their patients and flourish through providing deep compassion in all their interactions. They are committed to continually learning from their patients and families, their coworkers, and others who advocate for safe and comfortable dying, self-determined life closure, and effective grieving (NHPCO, 2012).

Sage/Master/Elder mystery watchers allow situations to guide their practice as transmitters of knowledge, relishing the opportunity to coach competent beginners, emerging apprentices, and skilled artisans from proficiency to expert, and beyond. This phase of hospice practice is characterized by the soulful use of Self, self-mastery with dignity and respect for self and others.

Mystery-Model Workplace

After all these years I have begun to wonder if the secret of living well is not in having all the answers, but in pursuing unanswerable questions in good company.
—Rachel Naomi Remen

The development of mystery watchers is a magical progression from fearful anticipation to serene competence. No matter how skilled an entry-level hospice practitioner may be, he or she will be challenged by the continual struggle to make sense of, and derive meaning from, the human suffering that accompanies his or her daily work. Practitioners' progression is dependent on so many variables that we have purposefully avoided assigning time frames to each phase of development. A crucial component of effective adaptation to a career in which death is a constant companion is the provision of a super-supportive work environment designed to bring out the best in each and every hospice practitioner. This environment will encourage the safe progression, although not necessarily linear progression, from one developmental phase to another.

Characteristics of a mystery-model hospice work environment include:

- hiring practices that reflect the importance of recruiting personnel with the potential for spiritual as well as professional growth;
- conducting death-awareness exercises during orientation in addition to teaching clinical palliative care skills;
- an expectation of intentional self-awareness of the stressors that may negatively impact work performance;
- an expectation of intentional awareness of the same in coworkers;
- resilience training to minimize the potential for burnout;
- staff support through consistent, professionally facilitated, and regularly scheduled support groups and individual counseling as needed;
- structured coaching based on appropriate and timely inquiry leading to a path of self- discovery;
- systematic mentoring that taps into current skills and attitudes and challenges forward movement through adaptive changes in work behavior;
- respectful supervision to guide deliberate and purposeful progression through developmental phases;
- mindful clinical supervision, promoting reflective, authentic, and compassionate practice;
- a high value placed on practitioners' and managers' self-care as a staff-retention strategy;

- resonant leadership that clears the barriers to hospice team members' success; and
- a rewards system that is unrelated to performance appraisal, avoiding the conflict of supervisor as counselor *and* judge over salary—a system in which people are paid fairly and equitably, even generously, for their position and in which nonmonetary rewards are valued.

With these characteristics in place, a culture of mutual support can flourish, in which reflective practice becomes accepted as the vehicle through which practitioners may constantly update their professional skills. Reflective practice is defined as reviewing practice experiences so that they may be described, analyzed, and evaluated, and consequently inform and change future practice (Bulman, 2008). This retrospective critical thinking exercise requires courage and openness on the part of all practitioners and those with whom they share the reconstruction of their experiences. While it offers opportunities to recognize strengths and weaknesses in the practitioner, it must always celebrate the valuable contributions that are made and must not become part of a formal performance appraisal. A mystery-model workplace establishes a standard for reflective practice, training coaches, mentors, or supervisors in the skills of facilitation of the process and providing not only the time but also the resources (expert facilitation, tools, methods, and guidelines) for the implementation of systematic reflective practice. The skilled facilitator encourages each practitioner to elicit feedback from peers and become a participant observer in the experiences of patients and their families. The model below, adapted from Johns and Freshwater (2009), identifies particular areas of interest commonly discovered through guided retrospective reflection.

Facilitated reflection on an experience of significance to the mystery watcher will result in:

- identification of personal issues arising from the experience, such as feelings, memories, and so on;
- clarifying personal/professional intentions;

- empathizing with patient/family and others involved in the experience;
- recognizing personal values and beliefs;
- comparing/contrasting this experience with previous experiences to create new options for future practice; and
- improving collaboration with other team members, patients, and families to meet assessed needs.

In-the-moment reflection is the basis of a mindful practice through which mystery watchers become aware of their thoughts and feelings within an actual patient/family experience. As the experience unfolds, mystery watchers question their interpretations and responses with the intention of ensuring the best possible outcome. Both retrospective and in-the-moment reflective processes are crucial in validating and legitimizing the kind of knowledge that is derived solely from the realities of hospice practice and plays such an important part in professional development.

To be effective, a mystery-model workplace must be managed by leaders who are in tune with themselves and with the people with whom they work. Boyatzis and McKee (2005) call such people resonant leaders, leaders who, through mindfulness, exhibit compassion and hope in the workplace.

A mystery-model workplace creates a healthy organizational culture that not only fosters well-being and safety at work but also focuses on work-life fit, the value of meaningful work and spirit in the workplace. Spirit at work is a distinct, multidimensional experience characterized by cognitive, spiritual, interpersonal, and mystical dimensions (www.kaizensolutions.org). A mystery-model workplace is founded on the belief that practitioners are engaged in meaningful work that has a higher purpose. There is an intentional awareness of the alignment between values, belief, and work. In addition, there is a sense of connection, through end-of-life care, to something larger than self. The mystery-model hospice interdisciplinary team engenders feelings of connectedness and common purpose that may become mystical experiences for those who are open to the transcendent nature of the extraordinary work. This type of workplace fosters the development of personal accountability for engagement, for excellent outcomes, for job satisfaction, and for spiritual growth.

Mystery-model workplace leaders understand that these times call for soul-based leadership, making the leaders available to transform negative workplace experiences with positive ones. Developing mutual respect and strong bonds among team members creates a work environment in which civility is honored and cooperative work relationships flourish (Lorenz, 2012).

A mystery-model workplace is founded on the recruitment of the right people, many of who come to hospice with their own mystery story clearly implanted in their soul. Effective mystery leaders develop a process of behavior-based interviewing, often including peer interviews, that endeavors to elicit a candidate's readiness to grow spiritually as well as a capacity to be clinically competent. Evaluating a candidate's communication and relational skills goes a long way in assuring a good fit with the hospice team.

Mystery-Model Workplace Tools

Everyone benefits personally and financially, including our communities and our nation, when courageous leaders advocate successfully for effective end-of-life care that includes the establishment of supportive environments in which the work of hospice can be accomplished. We include several tools for use in a workplace that values the development of mystery watchers.

The first tool below, “Interview Basics for Selecting Mystery Watchers,” provides guidance for one aspect of the recruitment process by outlining a format and work sheet for conducting candidate interviews.

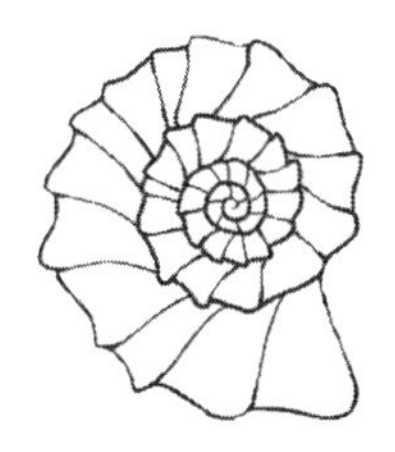

Interview Basics for Selecting Mystery Watchers

Interview Format

1. Introductions
2. Brief exploration of reason the candidate is interested in a hospice career
3. Brief discussion about the position
4. Interview questions (see below)
5. Candidate's opportunity to ask questions
6. Next steps

Interview Questions Worksheet

Select questions to address the candidate's knowledge, skills, behavioral traits and understanding of core competencies related to the mystery path. Create standardized questions including choices from samples below.

- Describe a time on any job when you were faced with stresses that tested your coping skills. What did you do?
- Relate a time in which you had to use your verbal communication skills in order to get an important point across.
- Describe a job experience in which you had to speak up to be sure that other people knew what you thought or felt.
- Describe a situation in which you felt it necessary to be very attentive and vigilant to your environment.
- Tell me a time in which you felt it was necessary to change your actions in order to respond to the needs of another person.

- What did you do in your last job to contribute toward a teamwork environment? Be specific.
- Describe a situation in which you were aware of the spiritual dimension of a person's suffering.
- Give some examples of ways you minimize stress in your life.
- Tell me about a time when your supervisor was not satisfied with the quality of your work. What actions did you take?
- Tell me a time when you were on a team and one of your teammates was not pulling his or her weight. How did you handle it?
- Share an example of how you were able to motivate employees or coworkers.
- Have you ever made a mistake on a job? How did you handle it?
- Tell me about a time when you worked effectively under pressure.
- Tell me about the part intuition plays in your efforts to problem solve.

****

A mystery-model workplace fosters management practices and strategies that value providing guidance about appropriate self-care activities for practitioners. We offer two tools as examples to implement this value. One is a self-care plan and the other a plan for resilience, both of which are implemented by the mystery watcher and monitored periodically by the mystery model leader.

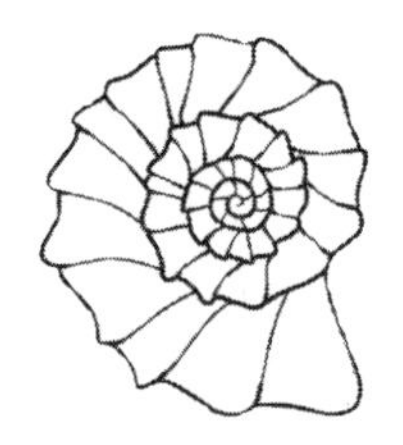

Mystery Path Self-Care Plan

Adapted from the work of Joan Halifax (2008), who asserts that keeping your personal life together is not an optional indulgence but an absolute necessity when it comes to being of use to others in the world. Mystery watchers develop a personal care plan to maximize their clinical effectiveness regardless of where they are on the mystery path.

Name of Mystery Watcher__________ Name of Buddy ____________

Name of Mystery Coach ____________ Date of Plan _____–_____

Directions:

- Circle stage on mystery path.
- Choose a practice for each aspect of self-appropriate to your developmental phase.
- Choose a buddy with whom to share your plan and who will help you to stick to it.
- Share the plan with your mystery coach; review/revise at least every three months.
- Keep a copy at work and at home.

	Novice	**Apprentice**	**Artisan**	**Sage/Master/ Elder**
Body				
Mind				

Spirit				
Psyche				
Social				

Practice examples:

	Choose self-care options with time frames, frequency, duration, and so on.
Body	Deep breathing, healthy diet, yoga, walk to work, rituals, keep on schedule, de-clutter
Mind	Meditation, study, less TV, read for pleasure, music, affirmations
Spirit	Mindfulness, explore retreats, attend a faith community
Psyche	Join a support group, consider a counselor, monitor stressors, set boundaries
Social	Have more fun, frequent date night with partner

"The more peaceful and accepting caregivers are, the more helpful we can be to the dying" (Halifax, 2008).

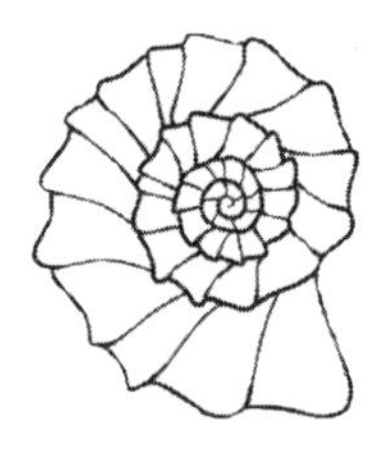

Plan for Sustaining Resilience

A mystery-model workplace (MMW) strives to minimize burnout by providing opportunities for professional maturation in the face of the potential for compassion fatigue. Each team member develops a personal plan based on the elements below. The plan is reviewed with a mystery coach monthly during the probationary period and at least every three months after that. The mystery coach provides feedback appropriate to the core competency level currently achieved.

1. Self-Care—aerobic exercise, evolve spirituality, meditate/pray, sleep well, eat well, be creative, relax, refuel, revitalize
2. Self-Regulation—self-awareness, scan body for muscle tension, mindfulness, monitor habitual perceptions, shift immediately to healthier response to stressor
3. Intentionality—follow personal/professional mission/code of honor, bring into alignment transgressions (even small ones), principle-based rather than demand-driven clinical practice
4. Perceptual Maturation—attend to what you can control and accept the rest, be at choice, self-validated, not in danger, potency balanced by humility
5. Connection/Support—develop/utilize workplace support, parallel charting, sharing narratives, sacred space/sanctuary

Adapted from J. Eric Gentry, PhD (2014).

"Between stimulus and response there is a space. In that space is our power to choose our response. In our response lies our growth and our freedom." —Victor Frankl

****

Reflective practice is truly the cornerstone of the acquisition of core competencies for reclaiming the mystery. The tool below provides a nine-step model that mystery-model leaders use to co-create reflective practice for the aspiring mystery watcher and themselves.

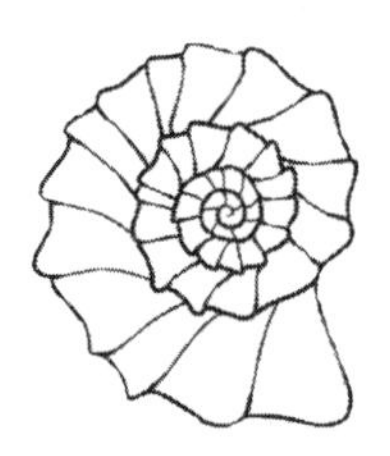

Mindful Clinical Supervision Process: A Nine-Step Model

The guidelines below assist the supervisor to create optimal retrospective reflection by the aspiring mystery watcher. It is important that the supervisor is trained to recognize when expressions of grief and other powerful feelings inherent in this process require professional counseling rather than workplace supervision.

1	Preparation	Creating the best possible environment for successful supervision, including location and timing; adopting a respectful approach
2	Security	Ensuring continuity of supervision from session to session, scheduling sessions, keeping the appointments without interruptions, maintaining confidentiality
3	Listening	Actively listening, with the intent to draw out significance from the practitioner's story
4	Clarifying	Ensuring the practitioner's perspective is heard correctly, guiding the mystery watcher to draw out significant issues, providing appropriate feedback
5	Understanding	Enabling the practitioner to gain insight into why he or she feels, thinks, and responds as he or she does in a particular situation
6	Options	Enabling the practitioner to identify and explore other, perhaps more appropriate, ways to respond in future situations and their consequences

7	Taking Action	Guiding the practitioner to draw conclusions and make choices for interventions in a future, similar situation
8	Empowering	Infusing the practitioner with the courage to act on insights, intuition, and profound alternative ways of knowing
9	Conclusion	Ensuring the practitioner has summarized the session, feels good about the lessons learned, and schedules follow-through as necessary

Adapted from Christopher Johns, *Becoming a Reflective Practitioner*. Second edition. (2004). Blackwell Publishing: Oxford UK.

****

Our final tool, "Mystery Watcher Self-Assessment Work Sheet," is designed to provide guidance and documentation as mystery watchers develop through the phases from novice to sage/master/elder. It may become the source document for reflective practice, providing the opportunity for mystery watcher and mystery model leader to measure progress toward wisdom and expertise. We introduce this tool with a testimonial from an aspiring mystery watcher who benefited from using this tool.

"As a new hospice nurse, I found the program an enriching experience. My previous nursing experience led me to try to be a fixer and problem solver of everything, even if most problems were unfixable for the very sick. The training I received helped me shift my perspective and my practice. During my tenure, I often became very frustrated by the demands of the job and distressed by the helplessness that comes with taking care of the dying. My feelings were causing me to lose focus on the purpose of my path. The weekly sessions helped me explore my own fears and setbacks in an honest and nurturing environment. With her (Lovvorn's) guidance, I feel I was able to positively refocus my energy toward my patients, and stand witness to the mystery of death. With my anxieties diminished, I feel I was able to better provide solace for my patients by simply being present, and listening quietly." —Caroline Gribbon RN, BSN, Chesapeake, Virginia, 2013

****

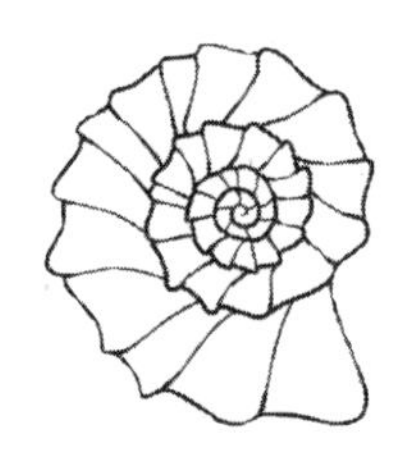

Mystery Watcher Self-Assessment Work Sheet

Practitioner: ______________________ Discipline: ______

Initial Phase: ________ Date: _________

Phases	Reflective Practice Notes – Self-Disclosure
N = Novice	
AP = Apprentice	
A = Artisan	
SME = Sage/Master/Elder	

Core Competencies	Phase	Dates Discussed	Coach/Mentor Comments/Plan	Initials
Sustaining Self	ALL			
Moving Through Fear	ALL			
Connecting	ALL			
Centering	AP–SME			
Comforting	AP–SME			
Collaborating	AP–SME			
Building on Strengths	A–SME			
Guiding Good-Byes	A–SME			

Honoring Truth	A–SME			
Transforming Knowledge	A–SME			
Standing By & Standing Firm	A–SME			
Accepting the Mystery	A–SME			
Notes:				

Coach/Mentor Name:____

Initials:________ Date Completed: ____________

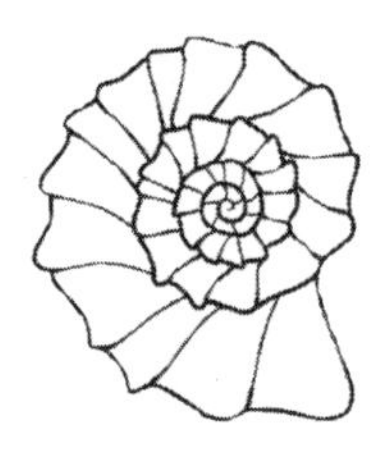

Epilogue

Nan Ottenritter, our friend and poet, graciously allowed us to use her "open letter to hospice" as the epilogue. We are forever grateful.

This Season Is Yours: A Thought and a Thank You

I am uniquely American.

Uniquely American means having frontiers to explore—wests that young men can head to—expanses of time and space. Modern physics suits us well as black holes open to the here and there of reality.

Uniquely American also means denying the ends of expanses—particularly the ends of our lives. Yes, the end of life can be viewed solely as a transition—but even transitions have an end before the change to a new normal, a new Hegelian center with the pendulum finally at rest. With elements of the left and right, it still centers at uniqueness. Every time.

What does this have to do with hospice? Everything.

As that uniquely American person, I strive, acquire, futurize, and keep my eyes on the prize. In so doing, I miss the prize at my feet, the joy of the moment, the life in the detail. I remind myself to:

- Learn from the ancients to slow down, be, observe.
- Learn from my aging that slowing is nature's way of bringing me to the moment.
- Simply stop.

What does this have to do with hospice? Everything.

I will tell you that I will be sad at my dying. Sad—because this life is so delicious, all about the senses. Frogs peeping in the early spring.

Chocolate and raspberries. Monet's Water Lilies. Vivaldi's Spring. Cool, rainbowed mist at the Cape of Good Hope. An absolutely tasteless joke. It is all so delicious.

What does this have to do with hospice? Everything.

Jean Shinoda Bolen would say that living with a serious illness brings us "close to the bone." People with cancer declare it was one of the best things that happened to them—a real wake-up call. Others say we are not afraid of dying but, instead, afraid of having not lived.

What does this has to do with hospice? Everything.

This season is yours, hospice. As Americans shift from exploration and acquisition to a more immediate and delicious life, they need help. I need help. If that help comes on the heels of a terminal diagnosis, so be it. If it comes at the deathbed of a loved one, so be it. If it comes late and long, so be it. It comes. And you are here.

No one is better equipped to help us than you, hospice. The stillness against which we rail is so needed. The deep listening soothes the deepest of pain. The simple not-aloneness makes dying, in the words of Griefwalker, a social event. It is at this time that we need the closeness of the tribe—whether we can articulate that or not. And you are here.

This season is yours, hospice. The immediacy of death brings life to … well … life. Deliciousness abounds. We are, at last, reconciled and even happy to be here—happy to be here as we transition to being there. And you are here—with us.

This season is yours, hospice. It is a holy calling, of that I have no doubt. It is a sacred space in which you work.

This season is yours, hospice—so you can help make our lives *ours* before that uniqueness melds into something unknown, that new unique center. You are here. Know that I am so thankful for that—and you.

By Nan Ottenritter, April 7, 2014

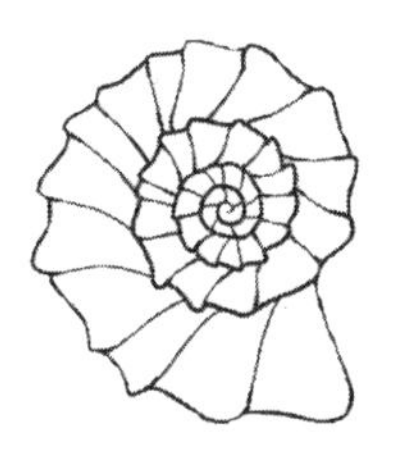

ACKNOWLEDGEMENTS

We are grateful for the help we have received from so many people as we have been writing this book, and most especially to Cory Cox, for his drawings and to Nan Ottenritter for her words.

The gratitude we feel for the hundreds of hospice patients and families who have informed our practice and from whom we have learned so much is impossible to over state.

And also to the many, many hospice team members across the country who have taught us so much; thank you. You are truly Mystery Watchers.

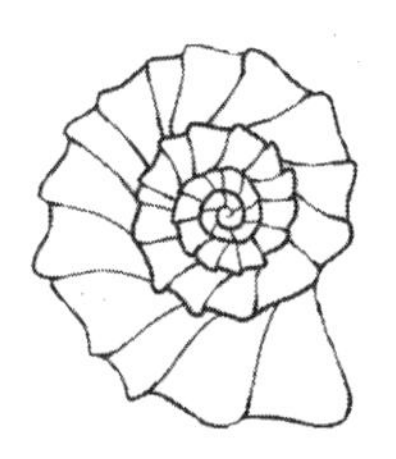

References

Balboni, Tracy A., Paulk, Mary E., Balboni, Michael J., Phelps, Andrea C., Loggers, Elizabeth Trice, Wright, Alexia A., Block, Susan D., Lewis, Eldrin F., Peteet, John R., and Prigeson, Holly Gwen. 2010. "Provision of Spiritual Care to Patients with Advanced Cancer: Associations with Medical Care and Quality of Life Near Death." *Journal of Clinical Oncology*. 28: 3: 445-52.

Benner, Patricia. 1984. *From Novice to Expert: Excellence and Power in Clinical Nursing Practice*. Menlo Park, CA: Addison-Wesley.

Boyatzis, Richard E. and McKee, Annie. 2005. *Resonant Leadership*. Boston, MA: Harvard Business School Press.

Brown, Brené. 2010. *The Gifts of Imperfection*. Center City, MN: Hazeldon.

Bulman, Chris. 2008. *Reflective Practice in Nursing*. 4th edition. Chichester, UK: Blackwell Publishing Ltd.

Cacioppo, John. T., and Patrick, William. 2008. *Loneliness*. New York: Norton & Co.

Callanan, Maggie and Kelley, Patricia. 1992. *Final Gifts: Understanding the Special Awareness, Needs and Communications of the Dying*. New York: Bantam Books.

Chopra, Deepak. 2010. *The Soul of Leadership: Unlocking Your Potential for Greatness*. New York: Random House, Inc.

Centers for Medicare and Medicaid Services (CMS). 2014. Presented at the "NHPCO's 29th Management & Leadership Conference: Leading and Mobilizing Social Change for 40 Years", March 25–29, 2014 at the National Harbor in Maryland. Accessed from the Internet June 20, 2014. https://www.cms.gov/Medicare/Medicare-Fee-for-Service-Payment/Hospice/Downloads/March-2014-NHPCO-Slides.pdf

Dictionary.com. 2014. Accessed from the Internet July 18, 2014. http://dictionary.reference.com/browse/Competence?s=t .

Dowling Singh, Kathleen. 1998. *The Grace in Dying.* New York, NY: HarperCollins.

Dreyfus, Stuart E. 1981. "Formal Models vs. Human Situational Understanding: Inherent Limitations on the Modeling of Business Expertise." USAF Office of Scientific Research, ref F49620-79-C-0063.

Erickson, H. Lynn. Editor. 2006. *Modeling and Role-Modeling: A View From the Client's World.* Cedar Park, TX: Unicorns Unlimited.

Ferrell, Betty. 2012. "Forgiveness in Palliative Nursing." *J Hosp Palliat Nurs.*14:8:501.

Ferrini, Paul. 1994. *Love Without Conditions.* Greenfield, MA: Heartways Press.

Gentry, J. Eric. 2014. *Compassion Fatigue Prevention & Resiliency.* Accessed from the Internet July 13, 2014. https://programs.gha.org/Portals/5/documents/societies/GONL/2012/Compassion%20Fatigue%20(GENTRY).pdf.

Goleman, Daniel. 2006. *Social Intelligence: The New Science of Human Relationships.* New York: Random House.

Halifax, Joan. 2008. *Being with Dying: Cultivating Compassion and Fearlessness in the Presence of Death.* Boston, MA: Shambala Publications, Inc.

Harper, Bernice C. 1977. *Death: The Coping Mechanism of the Health Professional.* Greenville, SC: Southeastern University Press, Inc.

Johns, Christopher & Freshwater, Dawn. eds. (2009). *Transforming Nursing Through Reflective Practice.* New York, NY: John Wiley & Sons.

Johns, Christopher. 2004. *Becoming a Reflective Practitioner.* 2nd Edition. Oxford UK: Blackwell Publishing.

Kearney, Michael K., Weininger, Radhule B., Vachon, Mary L.S., Harrison, Richard L., and Mount, Balfour M. 2009. "Self-care of Physicians Caring for Patients at the End of Life: Being Connected: A Key to My Survival." *JAMA.* March 18, 2009; 301:11:1155-64.

Labyak, Mary J. 2001. "The Experience Model". Reprinted in NHPCO's Membership Publication: *Newsline.* March 2012.

Lorenz, Joan. M. 2012. *Building Collegiality. Advance for Nurses*. Cited at www.advanceweb.com/nurses. May 7, 2012, pp. 25–28. Accessed from the Internet October 10, 2012.

Maslow, Abraham. 1970. *Motivation and Personality*, 2nd Edition. New York: Harper and Row.

Mead, Jonathan. 2014. *How to Use Gratitude to Re-center Yourself.* http://paidtoexist.com/how-to-use-gratitude-to-re-center-yourself/. Accessed from the Internet July 13, 2014.

Meditation 101: Grounding, Centering and Shielding. 2014. http://esotericaofleesburg.com/metaphysics/meditation_grounding.html. Accessed from the Internet July 13, 2014.

Myss, Caroline. 2002. *Self Esteem: Your Fundamental Power.* Boulder, CO: Sounds True. (CD)

National Healthcare Disparities Report, 2010. Accessed July 13, 2014. www.ahrq.gov/research/iomqrdrreport .

National Hospice and Palliative Care Organization (NHPCO). "2010 NHPCO Standards of Practice for Hospice Programs." http://www.nhpco.org/standards. Accessed October 10, 2012.

Orloff, Judith. 2014. *How to Center Yourself.* http://www.drjudithorloff.com/Free-Articles/Center-Yourself.htm. Accessed from the Internet July 13, 2014.

Palmer, Parker. J. 2011. *Healing the Heart of Democracy: The Courage to Create a Politics Worthy of the Human Spirit.* San Francisco, CA: Jossey-Bass.

Pearson, Patricia. 2004. *Opening Heaven's Door.* New York: Atria Books.

Peck, M. Scott. 1978. *The Road Less Traveled: A New Psychology of Love, Traditional Values and Spiritual Growth.* New York, NY: Touchstone.

Remen, Rachel Naomi. 2001. *My Grandfather's Blessings.* New York, NY: Riverhead Books.

Saunders, Cicely, Baines, M. and Dunlop, R. 1995. *Living with Dying: A Guide for Palliative Care.* Oxford, UK: Oxford University Press.

Tolle, Eckhart. 2005. *A New Earth: Awakening to Your Life's Purpose.* New York: Penguin Books.

Watson, Jean. 2008. *The Core Prinicipies/Practices: Evolving From Carative to Caritas.* http://watsoncaringscience.org/files/Cohort%206/

watsons-theory-of-human-caring-core-concepts-and-evolution-to-caritas-processes-handout.pdf . Accessed from the Internet June 3, 2014.

Winfrey, Oprah. "The Oprah Winfrey Show Finale." May 25, 2011. http://www.oprah.com/oprahshow/The-Oprah-Winfrey-Show-Finale_1/7 . Accessed from the Internet on October 11, 2012.

Zalenski, Robert J. & Raspa, Richard. 2006. "Maslow's Hierarchy of Needs: A Framework for Achieving Potential in Hospice." *Journal of Palliative Medicine.* 9.5.1120–1127.

Zerwekh, Joyce. V. 1995. "A Hospice Family Caregiving Model." *The Hospice Journal.* 10, 27–44.

About the Authors

Brenda Clarkson, RN, educated as a nurse in England, is now enjoying her fourth decade as a U.S. hospice professional. She is the 2015 recipient of the Hospice and Palliative Nurses Association's Vanguard Award in recognition of her pioneering efforts in certification of hospice nurses and her leadership in advancing the care of persons with serious illness across the nation as one of the first hospice surveyors for the Joint Commission.

Brenda is recognized as successful educator, presenting end-of-life topics at state, regional, and national professional meetings. She initiated the first efforts to improve end of life care in Virginia's prisons by bringing together professionals from hospice and prison healthcare.

Currently she is the executive director of the Virginia Association for Hospices & Palliative Care.

Myra L. Lovvorn, FNP-BC, is a national speaker focusing on end-of-life care. As a nurse, she concentrates on the spiritual aspects of living and dying. She is practicing the art of belief change through PSYCH-K® bringing integrative therapies into the medical community. She currently lives in Virginia, traveling and teaching holistic care of the living and the dying.

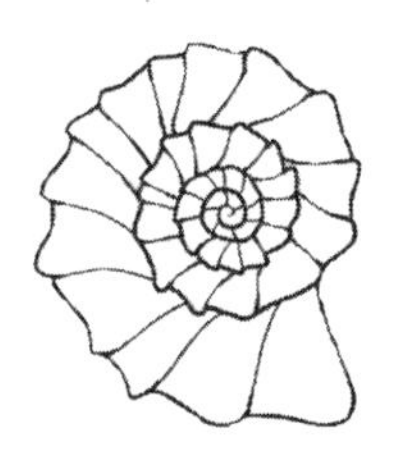

NOTES ON THE MYSTERY PATH

The following blank pages are included for the readers use in noting personal insights prompted by reading our book, examples from the readers' own practice and other reflections on the mystery path.